Origin of Life

By

Dr. Anab Whitehouse

© Anab Whitehouse
Interrogative Imperative Institute
Brewer, Maine
04412

All rights are reserved. Aside from uses that are in compliance with the 'Fair Usage' clause of the Copyright Act, no portion of this publication may be reproduced in any form without the express written permission of the publisher. Furthermore, no part of this book may be stored in a retrieval system, nor transmitted in any form or by any means – whether electronic, mechanical, photo-reproduction or otherwise – without authorization from the publisher.

Published 2018
Published by One Draft Publications
in conjunction with Bilquees Press

| Origin of Life |

Table of Contents

Introduction – page 7

Opening Remarks – page 13

What on Earth Is Happening? – page 23

Beach Front Property on a Warm Little Pond – page 45

Ah, Sweet Mysteries of Life – page 77

An Ocean of Difficulty – page 111

Monkeying Around with The Containment Blues – page 145

Science of Presumption Can Be a Beautiful Thing – page 181

Transposable Conceptual Elements – page 205

Closing Arguments – page 227

Introduction

What do you know about evolutionary theory? Or, maybe there are two questions here: (1) What do you think you know; (2) What do you actually know?

Quite irrespective of whether people believe in evolution or they are opposed to it -- most individuals probably would have to acknowledge that they know almost nothing at all about the actual nuts and bolts of the technical issues at the heart of evolutionary theory. Their beliefs concerning this matter -- whatever the character of those beliefs might be -- is, for the most part, rooted in two sources: (a) a largely unexamined acceptance of the opinion of others; (b) the extent to which evolutionary theory makes carrying on with the rest of their philosophical or religious perspective either easier or more difficult to continue to do.

In addition, the controversy surrounding evolutionary theory with respect to origin of life issues has been plagued by the fact that many of the advocates for various sides of this issue have been conducting the discussion on the wrong level. More specifically, people have been arguing mostly in terms of the evidence entailed by paleobiology ... that is, the anatomic/fossilized data that has been drawn from zoological and botanical studies. Unfortunately, the origin of life issue cannot be settled, one way or the other with any degree of certitude, when approached in this manner.

On the aforementioned level of discussion, one, at best, can obtain data that is either consistent with, or raises problems in, evolutionary theory as an explanation for the origin of life. However, there is no smoking gun (either for or against) to be found in such material -- just self-serving and heated rhetoric that tends to be cast in the garments of apparent rigor.

Furthermore, contrary to what many people believe, with the exception of a brief allusion to the possibilities that might exist in a 'warm little pond' somewhere ... a pond with just the right set of magical conditions ... Darwin has virtually nothing to say about the origin of life issue. The entire argument in his universally known but largely unread book is not about the origin of life but about the plausibility of a form of argument that alludes to, and presupposes, such a possibility without ever spelling out the mechanism.

The first part of the title of Darwin's historic work is: *On the Origin of Species by Means of Natural Selection*. There is a potential problem inherent in this title because the words tend to suggest that a species comes into being by a mechanism known as "natural selection". However, natural selection gives expression to a set of forces that operates after-the-fact of something having originated, and, therefore, at best, natural selection does not so much generate a species as much as natural selection operates on such a species once the latter has originated.

Natural selection acts on what is. It presupposes what is.

Natural selection does not cause what is, but, rather, it is an expression of those aspects of what is that might help determine which features of what is might continue to be. Natural selection introduces nothing new into the evolutionary picture, but, rather, the idea of natural selection only says something about the facets of that picture which might be most consonant with the dynamic of interacting natural forces existing at a given time and in a given location.

Therefore, the cause of that (whether a prebiotic collection of organic molecules or some primitive form of protocell) which natural selection comes to act upon still stands in need of an explanation. One cannot use natural selection as an explanation for that which natural explanation clearly presupposes without becoming entangled in completely circular thinking, and this sort of jaunt around the conceptual barn does not constitute an explanation of any kind.

Another problem with the previously noted title of Darwin's book is that it gives the impression that something is being selected ... as a person might make a selection among an array of choices. In truth, nothing is being selected since what exists in the way of a set of organic chemicals, or a set of protocells, or a set of species is either compatible (across a range of being more, or less, compatible) with the existing conditions of nature, or such chemicals, protocells, or species are not compatible. If random, such natural events do not select or choose.

What is compatible with the prevailing forces and conditions, survives. What is not so compatible tends not to survive. Nothing has been selected.

Another key idea in Darwinian theory is the notion of 'the accumulation of small variations'. The idea of the accumulation of small variations does not really account for either the origin of life, in general, or for the origin of different, particular biological blueprints, so to speak, on which the notion of species difference is based.

Variation presupposes that which is capable of such variation. Consequently, what needs to be explained is the origin of the capacity for variation.

Genetics is not the science that provides an account of the story of the origin of that capacity. Rather, genetics is merely a science that delineates how that kind of capacity operates once it has arisen.

Neither the ideas of natural selection nor variation help explain the origin of life. Only with the advent of modern molecular and cellular biology have we finally come into contact with the sort of information that allows one to make insightful judgments about the plausibility of evolutionary theory as an adequate account for the origins of life on Earth. When one integrates the disciplines of molecular and cellular biology with data derived from geology, hydrology, meteorology, and cosmology -- along with what has been learned about organic and inorganic chemistry -- then, one is in a position to work toward an informed understanding concerning the questions that surround and permeate the possibility of whether the modern neo-Darwinian theory of evolution offers an acceptable paradigm with which to approach origin of life issues.

In contradistinction to the original Scopes "Monkey" trial – when John Scopes, a high school science teacher, was put on trial for teaching material at odds with the Biblical account of the origins of man -- in *Evolution and the Origin of Life* Robert Corrigan, a fictional character, has been put on trial for teaching material that is considered by the book's prosecutor to be inconsistent with evolutionary theory. However, the defendant in this case is not a creationist nor is his argument an expression of what has come to be known as "Creationist Science".

The current overview is not about trying to prove the truth of this or that religious account of the origins of either human beings, in particular, or life, in general. *Evolution and the Origin of Life* is about the process of interpreting empirical evidence and subjecting that data to various methods of critical reflection.

Unlike works such as *Inherit the Wind* (which is largely the account of a clever lawyer's legalistic and philosophical dismantling of the simplistic arguments of a rather flawed personality who desired to be regarded as a defender of the faith), *Evolution and the Origin of Life* addresses the issue of whether, or not, science, as presently understood, can be said to demonstrate the validity of evolutionary theory as an account about the origin of all life. As such, the present overview focuses on the issue of evolutionary theory itself and does not get sidetracked with irrelevant considerations ... however interesting these later twists and turns might be in purely human terms.

At this juncture, some people might wish to make the critical comment that the foregoing really has little to do with modern evolutionary theory. The latter is an elaboration upon the seminal ideas of Charles Darwin and, as a result, is sometimes referred to as neo-Darwinian thought. If one would like to critically explore modern evolutionary theory, then one must stay within the confines of the neo-Darwinian paradigm as it is.

If someone made this sort of a comment, I might say something along the following lines. If such an individual is saying that modern evolutionary thought has no explanation for the origin of life on Earth, then let this fact be known far and wide so that everyone will clearly understand that the theory of evolution has absolutely nothing to say about how life came to exist on the planet Earth, and I will accept that perspective. Moreover, with the exception of changing a little terminology here and there in the discussion that follows, the following critical exploration concerning origins of life still poses a challenge to modern, scientific understandings concerning the issue of the origin of life.

More often than not, however, when people speak about the origin of life from a scientific point of view, they tend to use the term "evolution" in a broader sense than did Darwin. More specifically, such people tend to convey the idea that however life came into being (on Earth ... or arose elsewhere and, then, was somehow transported to Earth -- perhaps through meteors), it did so through purely "natural" evolutionary processes that generated increasing complexity involving prebiotic/inorganic chemistry that was, then somehow, 'naturally' transformed, in some evolutionary manner, into biotic chemistry, out of which the first protocells emerged – that is, the first species of life, and, at this point, neo-Darwinian theory would become relevant.

Evolution and the Origin of Life is primarily a critical exploration of this broader, more inclusive sense of 'evolution'. However, there are a variety of ideas entailed by such a discussion that carry implications for neo-Darwinian thought concerning evolution as well.

There are things about *Evolution and the Origin of Life* that are true. First, it contains a lot of technical material. Secondly, everything that is necessary for understanding this material has been included within the context of the direct and cross- examinations that take place during the trial and, as such, it is a largely self-contained work.

However, this work is not the sort of discussion that one can rush through. As with anything else worth the effort -- and I believe this book is worth the effort – *Evolution and the Origin of Life* takes time to digest and appreciate.

If you are ready to make the commitment to attempt to come to grips with the essential issues of evolutionary theory, then *Evolution and the Origin of Life* is waiting to be read. Be the first kid on your block to actually know what one is talking about when the conversation turns to evolutionary theory in relation to the origin of life problem ... and the foregoing point actually brings us to a third thing, alluded to previously, about *Evolution and the Origin of Life* that is true.

More specifically, if an individual cannot grasp the point-counterpoint of the discussion in this book, then, one is not in a conceptual position to argue intelligibly or honestly either for, or against, evolutionary theory. Whatever one might have to say on such issues will be entirely derived from the opinions of others -- opinions that might, or might not, be correct but with respect to which one will have no direct, personal understanding, knowledge or insight.

Opening Remarks

Upon arrival in Chicago, I took one of the shuttle buses from the airport that made the rounds of different hotels in the downtown area. After getting off the bus at my destination, the Balmer House, I confirmed my reservation at the main desk, picked up my key card and proceeded to the assigned room on the twenty-first floor.

I spent about ten or fifteen minutes in the room unpacking. Once this task had been completed, I went downstairs in search of the symposium registration desk.

After the signing in requirements had been met, I picked up a brochure that listed the various lectures, panels, discussions and so on that had been scheduled for the symposium. I quickly perused the day's listings.

The only event that struck my fancy was a moot court session on evolutionary theory to be held on the fourth floor, beginning at 3:00 p.m., about twenty minutes from now. I decided to go and see what it was like.

I fully expected the worst. At the same time, I held out a certain amount of hope that there might be some degree of entertaining diversion to be derived from the trial.

The whole thing would be very trying, indeed, if the participants took themselves too seriously and lacked a sense of humor. Equally daunting was the prospect that few, if any, of the individuals taking part in the moot court might know anything about modern evolutionary theory.

Images of Spencer Tracy and Frederick March came to mind from *Inherit the Wind.* There had been a remake of the movie in which Jason Robards played a Clarence Darrow-like character to Kirk Douglas's version of William Jennings Bryan.

I had enjoyed both movies but always felt the cards had been stacked rather unfairly in the debate. The crux of the drama had not really focused on evolutionary theory per se but on a clever lawyer's dismantling of a simplistic presentation of a narrowly conceived religious position held by a somewhat flawed personality. Hopefully, the moot court session was not going to repeat the same mistake, except in reverse -- that is, to use a clever lawyer's debating tactics to defeat a simplistic presentation of evolutionary theory.

If done properly, the trial setting could provide a valuable opportunity for a good educational experience. I preferred not to think about what the result would be if things were done improperly.

I eventually found my way to the indicated room. When I walked through the doors, two things surprised me.

For some reason, I was expecting a relatively small venue ... perhaps from having seen too much of the stage settings for the old, pre-revival, Perry Mason television series. The room selected for the trial was quite large and had been set up like an actual court complete with a jury box, witness stand, lawyers' tables, a raised desk-like affair for the presiding magistrate, and a large area at the back of the court room for the audience.

The other feature that I found interesting was the size of the crowd. Nearly every seat was taken. I was lucky to find a vacant chair.

The members of the jury already were assembled in their seats. Individuals that were performing as lawyers were at their respective tables.

A door to the left and behind the judge's bench opened, and a diminutive, attractive, forty-something, black-robed, brown haired woman entered the hall. As she did, a court officer stood up and said: "Hear ye! Hear ye! Hear ye! All rise, Moot Court is now in session, the Honorable Justice Karen Arnsberger presiding over the matter of the people versus Wayne Robert Corrigan in the City of Chicago, in, and for, the County of Cook, on June 26, in the year of our Lord, 2009. Draw nigh, and ye shall be heard."

The court officer watched the judge settle into her chair. When he was satisfied, the man announced: "Please be seated."

As the Judge waited for the noise of the audience's seating dynamics to subside, she shuffled and re-arranged some of the papers before her. When relative quiet had returned to the room, she scanned the court and, then, said: "In accordance with agreements reached in chambers between the prosecution and defense concerning pre-trial motions filed on various aspects of the procedural format to be observed during the course of this trial, the following principles will be in effect:

"(1) Due to considerations of time, the prosecution and defense each will be entitled, if so desired, to call a maximum of two witnesses;

"(2) With the exception of certain provisions ... provisions that have been agreed to by all parties concerned -- standard rules of evidence will be in effect throughout these proceedings;

"(3) Prospective jurors has been polled by both the defense and prosecution prior to the start of this moot court session and jurors have been selected and impaneled on the basis of their perceived capacity to judge the matter before the court in a fair and impartial manner. During the selection process, both sides were given the right to challenge seven of the candidates without the need to show cause for dismissal;

"(4) Again, out of consideration for the time constraints under which we are operating, neither the defense nor the prosecution will be permitted the opportunity for redirect examination;

"(5) The decision of the jury shall be read in open session on the last day of the symposium."

Putting the paper down from which she had been reading, she addressed each of the lawyers: "Are these the conditions to which you have agreed?"

Both responded, almost simultaneously, but slightly out of synchronization: "So stipulated, Your Honor."

"Very well," she replied.

She shuffled through a few more papers and stopped when she found the desired document. "Mr. Corrigan, will you please stand."

After the defendant -- a curly-haired, freckled youngster who looked to be in his mid-twenties -- had arisen, Judge Arnsberger said: "Wayne Robert Corrigan, you are being accused of teaching material to students that is in direct conflict with the facts of evolution as well as with the principles and methods of science. How do you plead?"

"Not guilty, Your Honor," came the response.

"All right, Mr. Corrigan, you may sit down," she indicated. Turning to the lawyer for the prosecution, she asked: "Are the people ready to proceed, Mr. Mayfield?'

"The people are prepared, Your Honor," he informed her. Looking in the direction of the table for the defense, she asked: "Is the defense ready to proceed Mr. Tappin?"

"We are, Your Honor," he stated.

"Good," she asserted, "then, let us proceed with opening statements. Mr. Mayfield, you are up first, and, gentlemen, please remember the meter is ticking."

Pushing his chair back as he arose, the lawyer for the prosecution -- who looked, sounded, and acted like he came from a family of moneyed- gentry ... walked to a point in front of the jury box, about midway between the two ends. He placed his hands momentarily on the railing atop the three-foot partition that enclosed the jury area and briefly made eye contact with various jurors as he looked first to his right and then to his left, as he surveyed the members of the jury.

Removing his hands from the railing, he began to address the jury as he slowly walked back and forth along the front length of the boxed area. Every so often, he would stop and face the jurors in front of him and speak as if he were talking just to them.

"Ladies and gentlemen of the jury, some seventy-five years ago, a man by the name of John Scopes was placed on trial for teaching evolution to his students. He was accused of promulgating theories and ideas that ran contrary to established religious doctrines concerning the origins of human beings.

"Today, you are being asked to pass judgment on a case that, in many ways, is quite similar to the Scopes case, but with a major difference. The defendant, Mr. Corrigan, has been accused of teaching material that is contrary to the facts of evolution and in opposition to established principles, practices and methods of science.

"Personally, I find it very disheartening that just as we begin our journey into a new millennium, and some hundred and forty -plus years after the publication of Charles Darwin's classic study: *The Origin of Species by Natural Selection*, we find ourselves unable, apparently, to put this matter behind us. I consider this situation to be unsatisfactory because for nearly one hundred and forty years, there has been an exponential growth of data from many different fields of scientific endeavor, all of which points in one direction -- namely, that evolutionary theory has been demonstrated to be a valid, consistent, empirically grounded, rigorously examined and scientifically satisfying account of the origins not only of species but of life itself.

"To be sure, as is true in any area of scientific research, there are differences of opinion concerning the value and use of various kinds of methods, techniques, and interpretations in evolutionary theory. However, none of these differences has anything to do with bringing into fundamental question, nor are they capable of undermining or refuting, the shared understanding and agreement of scientists concerning the essential character of evolution.

"At the heart of evolutionary theory is one simple truth. The origin-of-life, the origin of species, the transition from one species to another, -- these all are completely explicable in terms of known natural principles and processes.

"In other words, the principles of physics, chemistry, cosmology, geology, meteorology, and climatology, when combined with a few simple ideas such as natural selection and variation, provide a definitive, exacting and sufficient framework through which to understand the origins of life along with the biological phenomena that such origins set in motion. In short, the dynamic interaction that results from the interfacing of the forces operating through these various principles and processes is all that is necessary to be able to provide an adequate account of why certain phenomena and forms, rather than other phenomena and forms, were selected to play crucial roles in the emergence and perpetuation of different life forms.

"To employ principles and forces beyond the natural realm is to violate what is known as Ockham's razor. This long-venerated tenet of scientific methodology advises us not to multiply assumptions or concepts beyond what is needed to adequately account for any given phenomenon.

"Translated into more modern language, Ockham's razor is really the law of parsimony.

"Keep things simple. Do not complicate matters unnecessarily.

"Evolutionary theory operates entirely within the purview of this law of parsimony. Indeed, as far as the issues surrounding the origins of life are concerned, evolutionary theory is the only account that operates in accordance with this fundamental principle of rigorous methodology.

"The Scopes trial was caught up in emotion, dogma, and cultural biases. These influences settled like a dense fog around the minds and hearts of the jury and made reaching a fair and impartial verdict on the issues of that case very difficult.

"As a result, John Scopes lost the case. He lost the case despite the fact that the overwhelming character of the trial evidence revealed through testimony as well as cross-examination demonstrated that the charges against the defendant were entirely without merit.

"You, the members of the jury, have been selected because of your stated willingness to rise above issues of emotion, dogma and cultural bias. You have been selected because of your commitment to render a free and impartial judgment in the matter before us based solely on considerations of facts, logic and reasonableness of deliberations.

"The prosecution intends to demonstrate, within the limits being imposed on this trial, that evolutionary theory has been established beyond any reasonable doubt. Consequently, anyone, in this day and age, who would teach material that stands in opposition to a theoretical framework that has been developed and agreed upon during the last one hundred and forty-plus years can only do so by denying the facts of the matter and by refusing to observe sound scientific practice and principles.

"This is precisely the violation of which Mr. Corrigan is being accused. If the prosecution is successful in the presentation of our case, as I believe we will be, then you, the women and men of this jury, will, beyond any reasonable doubt, find Mr. Corrigan guilty as charged."

Once again, the prosecutor briefly ran his eyes down the two rows of impaneled jurors, stopping here and there to engage the eyes of this or that juror. When he had finished, he said: "Ladies and gentlemen of the jury, I want to thank you for the careful attention that you have given to my opening remarks. I am confident you will give the same considered attention to the evidence governing the case before you."

Mr. Mayfield turned and went back to his table. As he sat down, one of his assistants whispered something in his ear.

Judge Arnsberger turned to the lawyer for the defense. "Surf's up, Mr. Tappin," she informed him.

Before getting up, he picked up one of the sheets from the tabletop, looked at it for a few seconds, and, then, put the paper back down. He continued to sit for another five or ten seconds, as if in thought, and, finally, quickly rose and made his way to the jury area.

In speech and manner, Mr. Tappin appeared to be the opposite of Mr. Mayfield. With the exception of his thinking processes, everything about the defense lawyer was casual, informal and laid back.

Like the lawyer for the prosecution, Mr. Tappin appeared to be in his early thirties. Like Mr. Mayfield, the defense lawyer was moderately handsome but in a rough and ready manner and, therefore, somewhat at odds with the prosecution lawyer's aura of urbane sophistication.

"Good afternoon, ladies and gentlemen of the jury," the defense lawyer began.

"Good afternoon" was the collective, somewhat mumbled response from the jurors.

"I would like to thank my learned adversary for the wisdom of his comments," Mr. Tappin stated. "With his well-known and respected capacity for conciseness, Mr. Mayfield's introductory statement has focused on the most important elements of this case.

"The legal matters before you are not about ... or at least, it should not be about ... emotion, dogma and cultural biases. On the other hand, this case is about facts, logic and reasonable deliberations.

"These proceedings will not be about evolutionary science versus what some adversaries of evolution refer to as 'creation science'. This is so because my client is not an advocate of creation science, nor is this what he teaches in his classroom.

"My client, Mr. Corrigan, does not find any philosophical, or even religious, inconsistency between the vast majority of the tenets of evolutionary biology and a belief in a Divine Being Who creates the material and physical world. Mr. Corrigan is willing to admit the plausibility, if not tenability, of a position that says that evolution is merely the manifest form of the means through which God creates physical/material reality.

"The nature of Mr. Corrigan's faith is not so feeble that it depends on presupposing a particular conception of creation that precludes the possibility of evolution. He doesn't have a vested interest or axe to grind in this respect.

"Mr. Corrigan's concerns lay elsewhere. He is worried about issues such as truth, proof, logical argument, understanding, explanation, interpretation, and the integrity of the exploratory process.

"The case of the defense will not be about whether the second law of thermodynamics is inconsistent with the theory of evolution. We are quite prepared to live with the entirety of thermodynamic theory, including the relatively recent work on the phenomenon of dissipative structures that, sometimes, arise under conditions in which a system is far from equilibrium.

"The defense will not involve any arguments about whether the fossil record does, or does not, create problems for evolutionary theory. In addition, we will not try to exploit the controversies surrounding punctuated equilibrium theories as a means of undermining the framework of evolutionary biology.

"The position of the defense does not depend on the raising of questions about the reliability of dating methods based on radioisotopes. Furthermore, we have no intention of trying to use to our advantage differences of opinion concerning the role that, say, lunar samples play in pinning down the time of events on Earth, or the way in which, for example, high temperatures can affect the significance and interpretation of Carbon[12] and Carbon[13] ratios as an indirect procedure for helping to establish the possible presence of life at a given period of time in the early history of the Earth.

"There will be no attempt by the defense to take quotes of noted evolutionary scientists out of context and try to use these quotes as evidence against evolutionary theory. We are only interested in taking a look at what the best science of our day has to say in support of the case for evolutionary theory with respect to origin of life issues.

"Ladies and gentlemen of the jury, so far, I have told you what the case for the defense will not be. I have not, yet, indicated what our case will be, so let me take this opportunity to rectify that omission.

"The contention of the defense is as follows. When closely examined, evolutionary theories concerning the origins of life consist of little more than a rather argumentative mixture of: questionable assumptions, speculative conjectures, problematic inferences, arbitrary interpolations or extrapolations, ambiguous evidence, and a wonderfully serendipitous confluence of events quite beyond the ability of science to demonstrate with any degree of plausibility except, perhaps, to the true believers among evolutionary theorists who are more in need of faith to prop up their theories concerning the origins of life than are many followers of religious traditions.

"The defense will be asking you, the members of the jury, not to be dazzled by the technical virtuosity of modern science. We will be asking you not to be intimidated by the use of technical terms.

"However, the defense will be asking you to keep in mind the importance of such basic, fundamental questions as: How? Where? When? What? and Why? In addition, the defense will be asking you not to shunt aside or marginalize the number of questions that go unanswered within the evolutionary perspective.

"The defense believes that if the members of jury are prepared to persist in asking simple questions along the lines we have indicated, and if you are willing to keep a running total of the questions that, after all is said and done, lack a satisfactory answer, you will arrive at one conclusion beyond any reasonable doubt. This conclusion is that my client, Wayne Corrigan, is not guilty of teaching material in conflict with either the facts of the matter at hand or with the methodological tenets and principles of scientific investigation.

"Ladies and gentlemen of the jury, I would like to thank you for your kind attention to my opening statement. I also would like to leave you with one suggestion.

"Pause for a few seconds, sit back, relax and take a few deep breaths. For, in approximately ten to twenty seconds, you might not get the opportunity to do so again until these proceedings have concluded.

"Thank you, again," Mr. Tappin stated and returned to his seat. A few jurors seemed to be following his suggestion.

Approximately fifteen seconds later, Judge Arnsberger announced: "The prosecution may call its first witness."

What on Earth is Happening?

"At this time, the prosecution calls upon Professor Alan Yardley," proclaimed Mr. Mayfield. As he uttered the name, he looked back toward the audience.

A tall, thin, bearded man -- who appeared to be in his late thirties or early forties -- stood up in the area where Mr. Mayfield was looking. The man strode to the witness stand and remained standing while the oath was administered by a court officer: "Do you promise to tell the truth, the whole truth and nothing but the truth, so help you God?"

The witness answered: "I do."

The court officer then informed him: "You may be seated." Once the witness was settled in his chair, the court officer said: "Will you state your name and address for the record, please."

"My name is Alan Ross Yardley," he replied. "I presently reside at One Finch Beak Road, Daphne Major, the Galapagos Islands."

The court officer returned to his seat. Mr. Mayfield approached the witness.

"Dr. Yardley," he requested, "will you state your current occupation and title."

"I hold the Charles Darwin Chair for Biological Sciences at the University of Galapagos," he responded. "I am a full professor and teach a variety of courses dealing with different facets of evolutionary biology."

"How long have you held your present position, Dr. Yardley?" Mr. Mayfield inquired.

"For seven years," Dr. Yardley answered.

"Professor," Mr. Mayfield said, "would you be kind enough to list your major publications."

Dr. Yardley was about to begin when the defense lawyer arose. "If it pleases the court, Your Honor," Mr. Tappin indicated, "in the interests of saving time, the defense is quite prepared to acknowledge the expertise of Professor Yardley in the field of evolutionary biology. His reputation as a first-rate scholar is recognized internationally, and we feel there is no need to go through the usual procedures for establishing expertise with respect to this witness."

"So noted," acknowledged Judge Arnsberger. "Thank you for expediting matters, Mr. Tappin."

The defense lawyer nodded his head and sat down. He began writing something on a piece of paper and, when finished, showed it to his assistant.

"Your Honor," the prosecutor said, "before beginning examination of my witness, I would like to introduce into evidence, at this time, the People's Exhibit, marked 'A'." While saying this, he had returned to his table, picked up a collection of material, checked its identity, and delivered the bundle of papers to Judge Arnsberger.

The Judge examined the papers briefly and made a few notations, presumably, in her own log of the trial. Having done so, she said: "You may proceed Mr. Mayfield."

Returning to his table, he picked up another, similar bundle and walked back to the witness. Handing the papers over to Dr. Yardley, the prosecutor said: "These papers, Professor, that have just been introduced into evidence as People's Exhibit 'A', constitute a detailed curriculum for the courses on evolutionary biology that are being taught by the defendant, Mr. Corrigan. Dr. Yardley, have you had a chance to study these papers prior to the beginning of this trial?"

The professor quickly worked his way through the pile of documents. "Yes, prior to the beginning of these proceedings, I have looked through this set of documents," confirmed the professor.

"What is your opinion, Professor Yardley, of the educational merit of these curriculum materials as far as the teaching of evolutionary biology is concerned?" the prosecutor inquired.

"Well, in certain ways," he asserted, "they appear to be reminiscent of the kind of material that is taught under the misleading title of creation science. And ..."

"Objection, Your Honor," Mr. Tappin blurted out.

"On what grounds?" Judge Arnsberger asked.

"Your Honor, as has been clearly stated in the defense's opening statement, Mr. Corrigan's position is not that of the so-called 'creation scientists'. Unless the prosecution demonstrates in what way the position of Mr. Corrigan is 'reminiscent' of the position of the creation scientists and unless the relevance of that reminiscence to the present case can be established, then, all references to creation science are really immaterial and irrelevant to these proceedings, as well as being quite prejudicial to the interests of my client."

"Mr. Mayfield," inquired Judge Arnsberger, "does the prosecution intend to provide the court with the sort of demonstrations and connections about which Mr. Tappin is concerned?"

"No, Your Honor," indicated the prosecutor.

"Very well," she said. "The objection of the counsel for defense is sustained, and the statement of the witness will be stricken from the records. You'll have to begin again, Mr. Mayfield."

Nodding his head in compliance with the directive, the prosecutor turned backed to the witness. "Professor Yardley, in the light of what has just transpired, how would you sum up your objections to the curriculum materials of Mr. Corrigan?"

"Perhaps," the professor began, "the most diplomatic way to state what is problematic about the content of Mr. Corrigan's course material is that it is consistently antagonistic toward the precepts, findings, conclusions, principles, orientation and general framework of the modern theory of evolutionary biology. In other words, Mr. Corrigan seems to want to debate and question issues and themes that, for the most part, have long been accepted as settled among the vast majority of scientists all over the world."

"Dr. Yardley, is this 'antagonistic' flavor of Mr. Corrigan's teaching material, only directed at specific aspects of evolutionary theory or is the tenor of his attitude more general in character?" the prosecutor asked.

"Quite general, I would say, but it is manifested in specific ways at virtually every level of evolutionary inquiry. For instance, Mr. Corrigan seems unwilling to accept much of what has been agreed upon with respect to issues involving prebiotic chemistry, or the origins of the first proto cells, or the emergence of prokaryotic and eukaryotic forms of life, as well as ..."

"Professor Yardley, I'm sorry for interrupting you," Mr. Mayfield apologized, "but three or four terms, in quick succession, have occurred in the testimony, and I feel they should be explained by you ... in a brief fashion if possible ... for the benefit of the jurors. Perhaps you could start with the term 'prebiotic'."

"Certainly," the professor said, "I would be most happy to do so. 'Prebiotic' chemistry refers to the study of all chemical processes, whether inorganic or organic, which are thought to have occurred prior to the appearance of biological or living systems on Earth.

"These prebiotic chemical systems are believed to have evolved over the course of millions of years, into, first, quite primitive cellular forms of life known as 'protocells'. Such protocells were, however, sufficiently developed to exhibit three properties.

"First, they contained some kind of membrane mechanism that provided a certain amount of protection for, as well as enclosed an area involving, a variety of chemical reactions necessary to sustain life on some minimal basis. Secondly, there would have had to be a method of metabolism that would permit the coupling of certain sources of energy with the building up and tearing down of chemical substances that result in the regulation of cell functioning and structure. Thirdly, such a protocell would need a means of storing and replicating information concerning the capabilities of the protocell that would enable the entity to reproduce itself and generate other protocells of a similar enough nature to be able to perpetuate the life cycle in future generations.

"Before proceeding, however, I should point out something. Among evolutionary biologists, as far as the issue of protocells is concerned, the aforementioned three m's ... that is, membranes, metabolism and memory ... might operate in ways that are quite different from what goes on in the current, modern life forms with which we are familiar, such as prokaryotes and eukaryotes.

"Prokaryotic forms of life consist of single-celled organisms in which the genetic material of such an organism is not enclosed by a true nucleus within the cell but, instead, floats freely in an area known as the nucleoid. By and large, most prokaryotes are one species or another of either bacteria or blue-green algae.

"Eukaryotic forms of life, on the other hand, include all those organisms whose cells contain a true nucleus, consisting of a bilayered or double membrane, which ropes off, so to speak, a roughly circular area within the cell that stores the genetic blueprints for the cell. These eukaryotic organisms might be either single-celled or multiple-celled in character and, for the most part, involve all forms of life other than the aforementioned bacteria or blue-green algae prokaryotic life forms."

"Thank you, Professor Yardley, for your very concise definitions of the technical terms," said Mr. Mayfield. "I'm sure we will be relying on this ability of yours quite a lot in the testimony that lies ahead of us.

"Dr. Yardley, you have indicated in your previous testimony that Mr. Corrigan's curriculum materials take exception with well-established and generally agreed upon issues and themes at virtually every level of evolutionary theory. Maybe the most effective way in which to proceed is to spend some time providing an overview for the members of the jury concerning the theoretical framework for modern evolutionary biology.

"In this manner we will be able to develop, hopefully, a much clearer understanding of that to which Mr. Corrigan stands in opposition. Moreover, in the process of coming to this understanding, you can provide evidence that, when contrasted with the material in Exhibit 'A', will demonstrate the truth of the allegations contained in the People's charges against Mr. Corrigan.

"Let's start, Professor Yardley, with first principles. Could you provide us with an outline of the currently accepted understanding of the formation of the Earth and what ensued from that as far as the conditions that are believed to have arisen to give expression to the prebiotic environment out of which life is said to have originated."

"Objection, Your Honor," Mr. Tappin asserted. "While the defense is willing to concede that Dr. Yardley has expertise in the specific area of evolutionary biology, we are not prepared to concede his expertise in areas of cosmology, meteorology, climatology or geophysics."

"Under other circumstances, Mr. Tappin," the judge indicated, "I might be inclined to agree with you. On the other hand, earlier on, you waived your right to establish the precise nature of the parameters within which the expertise on evolutionary biology falls.

"Furthermore, unless I am mistaken, Mr. Tappin, in your opening statement you seemed to indicate that in order to set the stage for the case of the defense, you wished to concentrate on what the science of our day claims to be the best version of the evidence in support of evolutionary theory. Why don't we give them a chance to stick their head into the lion's mouth before trying to lop it off?

"I'm going to allow the witness to answer this line of questioning. Objection overruled.

"However, Mr. Mayfield, let's understand what is being said here. I don't want you taking undue advantage of the latitude that is being extended to you by the court, or else I will step in and revoke your privileges in this regard. Have I made myself clear?"

"Like a Norwegian fiord, Your Honor," he acknowledged.

"Dr. Yardley," the prosecutor said, "let me rephrase, somewhat, my previous question to you. Among evolutionary biologists, what is the generally agreed upon understanding concerning the conditions prevailing on Earth during prebiotic times?"

"To properly answer your question in even a cursory manner," stated the professor, "one must understand that prebiotic times entail a number of different stages and kinds of interacting evolutionary forces. These include: the evolution of the solar system as it relates to planetary formation; geological evolution; atmospheric evolution; hydrological evolution of the physical character, distribution and effects of the waters of the Earth; together with chemical evolution, especially as this development relates to the generation of increasingly complex forms of hydrocarbons that are the bread and butter of organic chemistry.

"I'll try to give a brief overview of all but the last of these areas. The topic of chemical evolution will require considerably more time.

"Obviously, my brief account of the issues beyond the horizons of chemical evolution will be leaving out a great deal of detail. Nonetheless, I believe people will be able to grasp the character of the general picture that is being constructed.

"To begin with ... ahh! Mr. Mayfield ... it is all right that I proceed in this way isn't it?" he asked.

"Of course, Professor Yardley," the prosecutor confirmed. "If I feel any clarification is necessary, I'll be sure to intervene.

"Moreover, Professor, I realize some minimum degree of technical language and explanation will be necessary. However, while avoiding as much distortion and oversimplification as possible, if you could try to make your account as clear and succinct as possible, this would be greatly appreciated.

"This is probably asking the impossible of you. Nonetheless, I believe the more you are able to approach the 'impossible' as a limit, the more easily will the jurors understand the validity of the allegations being made against Mr. Corrigan."

Professor Yardley paused briefly, seemed to gather his thoughts, and began to speak. "At one point in the development of cosmological theory," he said, "scientists believed planets were formed by a very rapid gravitational collapse of interstellar dust clouds once, depending on circumstances, certain critical densities within those clouds had been achieved.

"Today, based in large measure on the findings of the Apollo space program's crater studies of the moon, most scientists have abandoned the foregoing theory and, now, believe in an accretion theory of planet formation. In other words, they believe planets come into being, not through gravitational collapse of dust clouds, but by gradually growing in size by means of a series of collisions with other objects of varying sizes.

"For example, one begins with specks of cosmic dust that collide with one another to form tiny particulates. Particulates collide with other particulates as well as cosmic dust to form larger, gravel-sized objects.

"This cosmic gravel, in turn, collides with cosmic dust, particulates and other gravel-sized objects to generate larger and larger objects. Eventually, something the size of a small planet, called a planetesimal, is produced, and, then, later, through continued collisions, objects that are the size of the moon, and, finally, the Earth, emerge.

"The process of planet formation might have required a hundred million years give or take a few hours. This period of primary formation and evolution of the Earth has been determined, on the basis of radioisotope studies of the rate of conversion of uranium to lead, to have been completed approximately 4.55 billion years ago.

"As the objects grow larger, then, relatively speaking, there are fewer and fewer large size objects running around in space with which to collide. Collisions, of course, do continue to occur. Nonetheless, the number of years between large-scale, or even moderate-scale, collisions begins to increase.

"At first, after the formation of a planet the size of Earth has taken place, the occurrence of collisions will be separated by periods of time lasting hundreds, followed by thousands, of years. Later, the interval between collisions will become hundreds of thousands of years and, then, millions, if not tens of millions of years.

"The last great collision on Earth was believed to have occurred some sixty-five million years ago at the Chicxulub crater, some 300 kilometers in diameter, near the northern tip of the Yucatan Peninsula. This collision is thought to have led, both directly and indirectly, to the extermination of many, if not most, species of life, including the dinosaurs, living on Earth at the end of the Cretaceous era.

"In any event, most evolutionary biologists are agreed that life on Earth probably could not reasonably have been thought to have had the opportunity to establish a firm foothold until the frequency of these collisions had declined to, at least, less than once every ten or twenty million years. The reason behind this thinking is that whenever objects big enough to create craters of diameters equal to, or greater than, say, 265 kilometers, collide with the Earth, they cause, among other things, a one hundred-degree Celsius, transient rise in the temperature of the Earth's atmosphere.

"This would cause obvious, destructive havoc with the vast majority of origin-of-life processes that might have been going on in a prebiotic environment on Earth. There must be, consequently, enough undisturbed breathing room, so to speak, within which biological organisms would have a plausible opportunity for emerging spontaneously through purely natural chemical and physical processes.

"Most of my colleagues set the lower limit of the relatively undisturbed breathing space time that is considered to be necessary to account, reasonably, for the origins of, say, the first protocells, to be around ten to twenty million years. Such intervals of cosmic quietude are not likely to have taken place on Earth prior to about 4.44 - 4.41 billion years ago.

"These kinds of calculation are based on statistical projections derived from radioactive dating of the cratered surfaces of the moon. For instance, if one assumes there will be a proportionate increase in the number and size of large impacts as one goes from the smaller surface area of the moon to the larger surface area of the Earth, then, scientists have concluded there were about 15-16 collisions on Earth that were larger than the ones that caused the largest of the moon craters, Imbrium. These collisions would have taken place at some point after 4.3 billion years ago.

"Since collisions do not take place in accordance with a fixed schedule, they are a stochastic or probability phenomenon. Therefore, if we take the 15 or 16, previously mentioned, large-sized collisions with Earth and average them out over a period of time, we would have to wait for all of these collisions to take place before we could begin to talk about conditions on Earth that were minimally conducive, as far as collision activity is concerned, to the origins of life in a prebiotic environment.

"The time at which the last of these large-scale collisions is believed to have occurred is somewhere between 4.3 and 3.8 billion years ago. We should begin to find traces of life somewhere in this time-frame, and, in fact, we do, but I'll come back to this."

Professor Yardley picked up a jug on a table near the witness stand and poured water into a small drinking glass. He took a long drink, finished the glass, replaced it on the table, and continued on.

"When, as a result of the gradual process of accretion, the Earth grew to roughly its present size, our planet was not considered by scientists to be a static, dead entity. In fact, there were several theories about, for example, the formation of the core of the Earth that have ramifications for theories concerning the origins of life.

"One theory, the older one, maintained that the Earth started out as a cold body. Its interior layers did not begin to heat up until hundreds of millions of years later when there had been a sufficient amount of heat generated by the radioactive decay of various elements in the Earth.

"Consequently, rather than sinking to the core early on in the formation of the planet, heavier elements, like iron, remained fairly close to the surface for many millions of years. Moreover, since iron tends to react with oxygen, this reaction would have severely restricted the amount of oxygen that could have combined with carbon to form an atmosphere consisting of large amounts of carbon dioxide.

"According to this theory, the volcanoes created by the thermal activity of the Earth's interior layers would have caused the spewing forth, or out-gassing, into the exterior regions of the planet, of large amounts of nitrogen and carbon that would combine with hydrogen. These reactions would have led to an atmosphere consisting, predominantly, of methane and ammonia.

"If, on the other hand, one subscribes to the collision or accretion theory of planet formation, as most modern researchers do, then, one comes up with a very different sequence of events than is painted by the older theory that started off with a cold Earth. According to the up - dated theory, the many violent collisions that were typical of the Earth's early years would have generated thermal conditions sufficient both to melt the interior regions of the Earth, as well as the heavy elements, like iron, which were on the surface.

"As a result, the interior of the Earth, some two to four hundred kilometers below the surface, would have formed what is known as a 'magma ocean'. Among other things, this 'ocean' would have underwritten the activity of volcanoes for millions of years and would have served as the 'sea' by means of which the plate tectonics of landmasses would have manifested themselves.

"In addition, the heavy metals, such as iron, would have sunk, in the form of a dense liquid, thereby differentiating the Earth, through the formation of a magnetic core, at a very early stage of the planet's evolution. Iron, consequently, would not have been available to react with oxygen as the old theory hypothesized, and, consequently, this would have cleared the way for oxygen and carbon to combine to form an atmosphere consisting, to a considerable degree, of carbon dioxide instead of the methane and ammonia called for by the previous model.

"Calculations involving the atmospheric-mantle ratios of two isotopes, argon^{40} and xenon^{129}, suggest that as much as 80-85 percent of the Earth's atmosphere probably was out-gassed in the initial million years of the existence of Earth as a planet-sized body. The remainder of the atmosphere was slowly out-gassed during the following 4.4 billion years.

"In addition to large quantities of carbon dioxide gas, there is believed to have been considerable amounts of nitrogen gas in the prebiotic atmosphere. Furthermore, although trace amounts of sulfur dioxide, methane and ammonia also are considered to have formed part of the early atmosphere of the Earth, no oxygen was believed to be present in the Archean era atmosphere that lasted from about 4.54 until roughly 2.5 billion years ago.

"This assertion concerning the relative absence of any oxygen content in the Archean era atmosphere has been backed up by a variety of studies. For instance, research has been done in relation to the stability of certain compounds such as uranium oxide and iron oxide, and these studies strongly suggest that the oxygen content of the Archean era atmosphere prior to two billion years ago appears to have been extremely low."

"Excuse me for interrupting, Dr. Yardley," the prosecuting attorney interjected, "could you, perhaps, explain the significance of the relative lack of free oxygen in the Archean era atmosphere?"

The professor nodded in acknowledgement of the request and said: "Essentially, free oxygen is highly reactive and tends to remove hydrogen atoms from any compounds it encounters. If free oxygen were present in the Archean era atmosphere with anywhere near the concentration of roughly 20 percent of our current atmosphere, the tendency of oxygen to oxidize or to take hydrogen from other compounds would interfere, in a fundamental way, with many important chemical reactions in a prebiotic environment.

"If one were attempting, as evolutionary biologists are, to account for the transition from simple hydrocarbons to the more complex forms of hydrocarbons that are necessary to the emergence of biological organisms through natural processes, the presence of substantial amounts of free oxygen would undermine one's efforts. If the Archean era had an oxidizing atmosphere, this would constitute a major theoretical problem for evolutionary biology.

"Fortunately, we are not faced with such a difficulty. As I suggested earlier, the available evidence indicates oxygen was not present during the Archean era except, at best, in minimal, trace amounts."

"Thank you, professor," Mr. Mayfield stated. "Please continue with your overview."

Dr. Yardley seemed to be searching in the air for where he had left off in the previous discussion. Apparently finding it, he said: "The process of core formation through the downward displacement of dense liquids consisting largely of molten iron is believed to have generated enough heat to raise Earth's temperature by as much as 1500 degrees Celsius. Such temperatures, in turn, could have helped create a set of conditions on the surface of the planet that might have culminated in a runaway greenhouse effect that, for a period of time, would have resulted in a melting of the surface of the Earth, creating a magma ocean of truly global proportions.

"This forms part of a theoretical scenario that is referred to as the 'hot world hypothesis'. A number of scientists have conjectured that, among other things, the Earth's crust would have been extremely thin during this period of geological evolution.

"These researchers believe that such a thin crust would have been very prone to cracking, and, one of the results of this would be the prevalence of a great many more hydrothermal vents than exist currently. These hydrothermal vents were channels to subterranean rivers and oceans of molten rock.

"Such hydrothermal vents would have helped create conditions for such phenomena as underwater geysers. In addition, they could have played an important role in providing a set of conditions out of which life might have first arisen.

"Modern researchers, however, also link the origin of the oceans and their concomitant hydrogen cycle with the previously mentioned process of out-gassing. Voluminous quantities of water would have been released by the heating of the Earth's mantle.

"This water vapor would have condensed, subsequently, into the extensive precipitation that formed the oceans. In addition, this process of condensation would have created a cooling trend that, eventually, would have helped to cool the atmosphere and surface of the planet down to the range of 40 degrees to 80 degrees Celsius that is believed to have prevailed at the time of the emergence of life from the prebiotic environment.

"In any event, most scientists agree this sequence of steps involving: A, the formation of the Earth's core, B, the gradual evolution and retaining of an atmosphere consisting of large amounts of carbon dioxide, C, the formation of oceans, as well as, D, the cooling down of the surface to temperatures in the range of, say, 40 to 80 degrees Celsius, was not likely to have been completed before 4.44 to 4.41 billion years ago, some eleven to fifteen million years after the emergence of the Earth as a planet-sized body. This figure coincides roughly with the evidence mentioned earlier concerning the gradual lessening of collisions with Earth of objects sufficiently large to interfere with, or frustrate, the prebiotic processes that eventually resulted in the formation of either protocells or biological organisms.

"There is further, independent data that helps confirm the foregoing time frame. These studies concern the mineral zircon.

"Zircon does not dissolve during the process of erosion. This mineral becomes deposited in sediment in the form of particles.

"Zircon particles are capable of lasting for billions of years. As such, zircon can provide evidence concerning the time of formation of a relatively stable surface crust.

"Ancient particles of this mineral have been found in Western Australia. These specimens were dated as having been in existence from around 4.1 to 4.3 billion years ago.

"The discovery and dating of these zircon particles is said to demonstrate there was a differentiated crust, consisting largely of silicon-derivatives, already in existence by that time. With the exception of various volcanic islands that had risen above sea levels, the aforementioned crust was believed to have been covered by a global ocean whose pH value is commonly set at 8.0, plus or minus 1 ... that is, this massive ocean was considered to have a pH that was either slightly basic in character or was relatively neutral.

"Among the oldest fossils discovered by scientists are structures known as stromatolites. Communities of marine microorganisms consisting mostly of cyanobacteria have produced those structures.

"Stromatolites are a combination of sedimentary material of various kinds that have been trapped in an inorganic secretion generated by these organisms. The ones that were produced at least 3.55 billion years ago are homologous with, or very similar in structure, character and appearance to the ones that are produced today.

"The oldest known stromatolite structures have been found in the lower strata of the Warrawoona Group of rock formations in Western Australia. This Group is the second oldest well-characterized rock formation that is known to scientists.

"The oldest such rock formation that, so far, has been encountered is the Isua Supracrustal Belt in Southwestern Greenland. This has been dated at about 3.77 billion years ago.

"The Isua formation consists of high-grade metamorphic rocks that have gone through a process of reformation under conditions of extremely high temperature and pressure. Consequently, any direct fossil evidence that might have been contained in this rock formation would have been destroyed.

"However, there is some indirect evidence that has been discovered at Isua to suggest a bacteria-like organism might have existed in Greenland some 3.85 billion years ago. This evidence is based on an analysis of the ratios of two isotopes of carbon, C^{12} and C^{13}, that were found in a hydrocarbon specimen taken from the rock formation.

"Since C^{12} tends to be used preferentially in biological processes rather than C^{13}, and since the ratio of C^{12} to C^{13} found in the sample of hydrocarbon was high, some scientists have been quite excited by the implications of the findings. They have concluded, despite possible methodological contraindications, that these findings on carbon isotope ratios might mean the hydrocarbons being examined were produced some 3.8 billion years ago during a process of photosynthesis in which an organism converted carbon dioxide into oxygen along with various hydrocarbon compounds.

"Interestingly, the term 'Isua' is translated from the Inuit language as being equivalent to the English phrase: 'the farthest we can go'. Whether this is true as far as the earliest evidence for life is concerned remains to be seen.

"Be this as it may, if the scientific interpretation of the significance of this analysis of the Isua hydrocarbon is correct, then, the earliest evidence for life has been placed just some 750 million years from the time the Earth reached planetary size. Furthermore, if the interpretation of the carbon isotope ratios is correct, living organisms have been located only 200 - 400 million years from the time when the prebiotic conditions on Earth are thought to have begun to stabilize with respect to a broad set of planetary, geological, atmospheric and hydrological parameters considered to have an important bearing on the issue of the origin-of-life.

"This period of 200-400 million years establishes the temporal framework within which modern evolutionary biology has attempted to delineate a plausible sequence of steps in chemical evolution. This sequence would provide an account of the dynamic of factors considered necessary to produce a working prototype of a living organism capable, minimally speaking, of processes of photosynthesis similar to what is suggested by the Isua hydrocarbon.

"Conceivably, there might have been some primitive form of life, a protocell, which existed prior to the emergence of the first modern prokaryotic-like microorganism. On the other hand, its manner of cellular functioning probably would have been very different from, and, therefore, a matter of speculation relative to, the kind of DNA-based organism that is indicated by the earliest evidence we possess either with respect to the indirect evidence of the Isua rock formation in Greenland or the direct evidence of the Warrawoona Group in Australia.

"On the basis of the available evidence, the Isua hydrocarbon and the Warrawoona prokaryotes constitute remnants of the last ancestor that is shared or held in common by all existing life forms. More distant or ancient ancestors, in the form of various kinds of primitive protocells, do not necessarily form part of the biological lineage of all current life forms.

"As such, these kinds of protocells would be regarded as spontaneously arising experiments in life that, for whatever conditions of natural selection, fizzled out at some point. These experimental failures, if you will, are to be distinguished from the appearance of the first, sustained, experimental biological success story to emerge from the prebiotic environment and that represents the last common ancestor of all subsequent life forms."

"All right, Dr. Yardley," the prosecutor said, "you have established a general framework within which, and through which, a person can engage the more difficult issues surrounding chemical evolution. For the benefit of the jurors, let's try to break up the themes of chemical evolution into units that, to the degree this can be accomplished, will become a little bit more user friendly for those of us who are relatively uninitiated in such matters.

"Professor, if you had to list four or five areas of discussion that you consider to be crucial to developing some minimal appreciation of how evolutionary biologists go about explaining the transition from prebiotic chemistry to the first life forms, what areas would you cite?"

Hesitating only slightly, Dr. Yardley replied: "First, one should address the ways in which more complex hydrocarbons either evolved out of chemical reactions amongst simple hydrocarbons or became available to the prebiotic environment through means other than chemical reactions. Secondly, there would have to be some discussion of the systems of energy that were helping to drive the chemical reactions in the prebiotic environment.

"At some point one would have to talk about the formation of proteins by the linking together of amino acids through peptide bonds. This would be of great importance because of the many different roles that proteins have in biological organisms, including: hormonal functioning; muscular contractions; the variability of morphology or structural form among species; electron transport in both photosynthesis and respiration; antibody activity in the immune system; and, the transport of nutrients, ions and so on across the membrane barrier.

"Quite obviously, one also would have to explore the processes surrounding the formation of nucleic acids, especially, of course, deoxyribonucleic acid or DNA and ribonucleic acid, RNA. Both of these molecules have fundamental roles to play in the processes of replication, transcription, translation and energy-coupling reactions that are central to the continued existence of both individual organisms as well as a given species.

"Finally, one would have to discuss the role that lipid formation plays in, for example, the structure and function of cell membranes. Biological membranes help regulate the passage of compounds into and out of the cell, and, in doing so, provide a relatively protected, enclosed environment in which various vital chemical reactions can take place under much more favorable conditions than might be prevailing in the medium that is surrounding the cell's exterior".

"In view of the limited time available to us, Dr. Yardley, I am hoping you will be able to summarize some of the research evidence concerning the different areas you have just mentioned that scientists believe helps establish a compelling case in support of the modern theory of evolution. In fact, Professor, maybe the easiest way to proceed is to allow our discussion to unfold in accordance with the sequence of topics you have listed.

"Consequently, if you will, Dr. Yardley, begin with the first theme you cited as being important to the foundations of modern evolutionary theory. This concerned, I believe, the generation and availability of complex hydrocarbons in the prebiotic environment."

"There are," the professor said, "two broad approaches to explaining the existence of complex hydrocarbons in the prebiotic environment. One approach focuses on the chemical reactions and dynamics that are likely to have occurred on the Earth in prebiotic times.

"The other approach, which is not necessarily in conflict with, or in opposition to, the first approach, gives emphasis to the possibility that various hydrocarbons, both simple and complex, might have been transported to Earth through carbonaceous chondrite meteors, comets and interplanetary dust particles. I'll start with this second approach.

"The term chondrite is derived from the millimeter-sized structures -- known as chondrules -- that can be found distributed throughout the interior matrix of a meteor along with other kinds of stony minerals. The origin of these chondrules still has not been determined although they are believed to come from the aggregates of silica minerals that were generated through the melting and fusion occurring in the solar nebula during the early stages of the evolution of our solar system.

"Approximately 5-6% of these stony, chondrite meteorites also contain different amounts of carbon compounds. For obvious reasons, this subset of stony meteorites is referred to as carbonaceous chondrites.

"Usually speaking, carbonaceous chondrite meteorites contain up to several percent, by mass, of carbon materials, of one sort or another. Moreover, some of these compounds include complex hydrocarbons.

"For example, the Murchison meteorite that fell in Australia in 1969 has been studied quite extensively. Six of the basic twenty amino acids found in Earth organisms were discovered in that meteorite.

"There also were at least twelve other kinds of amino acid compounds found in the meteorite. Although, as far as we know, these other varieties of amino acid do not occur in biological organisms on Earth, their presence is considered significant because it suggests, under the right prebiotic conditions, many different species of complex amino acids are capable of being formed.

"Some people have disputed the Murchison findings, claiming that the amino acids discovered in the meteorite were there as a result of contamination by organic matter from Earth. While most researchers do not accept such claims, there is a small aura of controversy lingering about the Murchison meteorite.

"This charge of contamination cannot be leveled at the findings of another study involving two meteorites that have been discovered in Antarctica. These meteorites had been buried in the frozen depths of Antarctica's ice for some 200,000 years.

"Many varieties of amino acid were found in those two meteorites. A little less than half of these amino acid compounds were quite different from the ones that are found in living organisms on Earth.

"The definitive proof concerning the extraterrestrial origin of these amino acids has to do with their optical properties. More specifically, by optical properties, I mean the direction in which a solution of such amino acids can rotate the plane of polarization of polarized light that is passed through such a solution.

"On Earth, when one shines polarized light through a solution of amino acids taken from a biological or living source, then, in such a solution, all twenty of the amino acids that form the proteins in Earth organisms will rotate, to the left, the plane of polarized light shining through the solution. This is a distinctive signature of the amino acids of Earth organisms.

"On the other hand, if one throws together a batch of amino acids in the laboratory, one will end up with what is called a racemic mixture. In other words, there will be equal numbers of what are called, in accordance with an agreed upon convention, left- and right-handed amino acids.

"This means that if one were to shine polarized light through solutions made up of this racemic mixture, one would find the direction of rotation of the plane of polarization shifting in different ways. Sometimes the direction of rotation would be to the left, and sometimes the shift in the plane of rotation would be to the right.

"When, however, amino acids from these meteorites were placed in solution, they shifted the plane of polarization exclusively to the right. This was entirely unlike what happens with either the racemic mixtures of amino acids in the laboratory or the amino acid solutions drawn from organisms on Earth.

"At least two conclusions follow from this. First, the only explanation we have for the origins of the amino acids in the Antarctic meteorites involves sources that are extraterrestrial in nature. Secondly, the existence of such complex hydrocarbons suggests that when conditions are right, whether on Earth or elsewhere, amino acids will arise through natural processes.

"In addition to amino acids, other kinds of complex compounds have been found in some carbonaceous chondrites. One researcher, for instance, discovered hydrocarbon compounds that appeared to have properties that could have played a role in membrane formation.

"This same researcher also found a yellowish pigment-like material that was able to absorb energy when light was shone on it. This pigment might have been some sort of precursor to, or an early competitor of, the chlorophyll pigment system that eventually emerged in some Earth organisms."

Professor Yardley paused in his presentation to pour another glass of water. Once he filled the glass, however, he did not drain the glass as he had done previously.

He held the glass in his hand and took only occasional sips. After one of the samplings, he said: "The material strength of carbonaceous chondrite meteorites often is so low many of them are unable to traverse the Earth's atmosphere without undergoing an airburst phenomenon in which they break up, and there is a release of many megatons of energy. Nonetheless, this sort of disintegration results in an increased surface-area-to-volume ratio of the remaining fragments that might allow some of the remnants to reach the ground with their organic payloads still intact.

"Researchers, in fact, have recovered fragments from catastrophic airbursts that are about a millimeter in size. Those who have examined such fragments have observed no signs of heating in their interiors and, therefore, any organic compounds that could have been there would have been protected from the effects of both the explosion as well as the heat of friction from passage through the Earth's atmosphere.

"Comets have been hypothesized, by some researchers, to be another potential means of transporting various kinds of hydrocarbons to Earth. These individuals have estimated -- on the basis of different methodological considerations -- that the composition of comets might have a hydrocarbon content which constitutes up to 14% of the mass of the comet.

"However, certain kinds of disparities between, on the one hand, the cratering records of the satellites of some of the outer-most planets, and, on the other hand, the cratering records of the so-called terrestrial planets that are closer to the sun, have led some scientists to maintain that very few comets are likely to have collided with Earth. Considerable uncertainty surrounds the role, or lack of it, which comets might have played in delivering organic molecules to Earth.

"There are some scientists who have argued that a far more important method of bringing organic compounds to the Earth might involve what are known as interplanetary dust particles. These particles, which might be the remnants of comets or asteroid-asteroid collisions, are about a micron in size, ... about one -thousandth of a millimeter.

"This might seem excessively small, but one should keep in mind, many bacteria are no more than one micron in diameter. Moreover, bacteria contain many, many, very complex hydrocarbon molecules.

"Not only are interplanetary dust particles big enough to contain, potentially, a variety of complex hydrocarbons, some of these particles might have just the right kind of mass properties that would prevent them from being incinerated by the frictional heat that is generated during entry into the Earth's atmosphere. Some researchers have calculated that those dust particles that are between: 10^{-12} to 10^{-6} grams, would be decelerated sufficiently in our atmosphere to allow such particles, which have been radiation-hardened by their trip through interplanetary space, to reach the surface intact.

"If the dust particles were smaller than this, they probably would be destroyed by the photolysis that is brought about by the ultraviolet part of the spectrum of sunlight. If, on the other hand, the dust particles were to approach the size of, say, small pebbles, they would be destroyed by organic pyrolysis, or the decomposition brought about by the heat of friction when traversing the Earth's atmosphere.

"Approximately 10% of an interplanetary dust particle's composition is in the form of hydrocarbon molecules. In addition, some individuals have estimated that the collective mass of the particles that enter our atmosphere outweighs many of the smaller, grapefruit -sized, meteorites by a ratio of approximately 100,000 to 1.

"Some researchers have calculated that carbonaceous chondrite meteorites and comets, when considered together, could have transported as much as 10^{20} grams of organic carbon, or hydrocarbons, to Earth during the prebiotic period that led to the origin-of-life through natural chemical processes. If one adds this amount to that which is believed to have come through interplanetary dust particles, then one is talking about quite a lot of organic carbon materials.

"Irrespective of the precise extraterrestrial or exogenous source of the hydrocarbons, evolutionary biologists believe these organic contents would have been released over time. Heavier, water - soluble compounds, like amino acids, would have dissolved in the global ocean.

"Low-density hydrocarbons, on the other hand, are likely to have become concentrated on the surface of the ocean ... much as an oil-slick does today.

Eventually, these molecules, like so much flotsam, would surf on the tides to the shores of volcanic islands or continents that were in the process of formation.

"The same mechanism of tidal transportation, of course, also would occur in relation to the heavier water-soluble compounds that went into solution in the ocean. The process probably just would have taken longer."

Mr. Mayfield was about to ask another question when a man came through the door behind, and to the left, of the judge. The man approached the judge and seemed to be relaying some message to her in the form of a folded piece of paper.

Judge Arnsberger took the paper silently and nodded her head in acknowledgement or thanks to the man. She scanned the piece of paper briefly, and, then, put it down.

"Mr. Mayfield," she said, "before you continue with your direct examination of this witness, I'm afraid there is an urgent matter that awaits me in chambers. I ask for your indulgence and extend my apologies, but I need to call a short recess often to fifteen minutes."

Having made her announcement, she banged her gavel. She quickly got up from her chair and soon disappeared behind the door through which the messenger recently had come.

Beach Front Property on a Warm Little Pond

The door at the front of the courtroom opened and the judge entered. A court officer said: "All rise," and, then, a short time later: "Please, be seated."

"You may continue with your examination of the witness, Mr. Mayfield," Judge Arnsberger directed. "I should remind the witness that he is still under oath."

"Dr. Yardley, I believe," indicated the prosecutor, "you were talking about meteorites and carbonaceous chondrites before the recess. Would you continue on with your testimony please?"

"Actually," the professor stated, "I was just about to begin talking about something else when the recess was announced. As I suggested earlier in my testimony, meteorites, comets and interplanetary dust particles are only one approach to explaining the presence of various kinds of hydrocarbons, both simple and complex, in the prebiotic environment of early Earth history. The other approach, to which I will now turn, concerns the chemical processes that are believed by evolutionary biologists to have been operating prior to, but which eventually brought about, the advent of biological organisms.

"Serious experimental work in the area of prebiotic chemistry has been going on for nearly fifty-five years in laboratories all over the world. Symposia and conferences dedicated to this subject take place on a regular basis, and, in addition, there are academic journals that publish articles dealing with virtually every facet of the prebiotic chemistry in which life is believed to have had its origins.

"Obviously, I cannot possibly present all of that material at this time. What I can do, however, is to try to provide some of the broad-brush strokes of the picture being painted by researchers.

"Although a few scientists, such as Alexander Oparin in the Soviet Union and J.B.S. Haldane in England, had been doing work on this topic during the 1930s, many people cite the early-1950s work of Stanley Miller and Harold Urey at the University of Chicago as marking the real beginning of serious investigation of the conditions necessary for the chemical origins of life. They were the first to put things to the test under laboratory conditions.

"In an oft-cited, classic experiment, Miller gathered some gases, such as methane (CH_4) and ammonia (NH_3), believed to be present in the early

Archean era atmosphere, subjected these gases to a continuous spark discharge, which was intended to simulate the action of lightning, and examined the results after a number of days. The laboratory procedure had generated a variety of amino acids, some of which are found in living organisms and some that are not present in life on Earth.

"Amino acids are complex hydrocarbons. They consist of three parts.

"One part is a carboxyl group, having a formula of COOH. A second component is an amino group with a formula of NH_2.

"The third aspect of the amino acid is a side chain. This varies, in a unique way, with each, different amino acid.

"Some 16-17 years after Miller's experiment, the Murchison meteorite was found in 1969, and scientists were able to demonstrate a number of similarities between the products of Miller's experiment and the hydrocarbons found in the meteorite. For instance, they discovered that the kind and quantities of amino acids found in the Miller experiment were very, very similar to the kind and quantities of amino acids found in the meteorite.

"In any case, by 1953, Miller had produced the first experimental evidence that natural chemical processes could produce complex organic compounds that are fundamental to life on Earth. Over the next forty-odd years many other experimental results would be forthcoming from Miller and other researchers.

"In one series of experiments, Miller and Urey discovered that roughly 10% of the carbon molecules contained in the gases of their experimental set-up eventually were converted into known organic compounds. Furthermore, as much as 2% of this converted carbon was involved in the generation of amino acids within the experimental apparatus."

The prosecutor, Mr. Mayfield, who had been listening intently to the professor, suddenly came to life, so to speak, and said: "Dr. Yardley, earlier you had indicated that an oxidizing atmosphere ... in other words, an atmosphere composed of, say, oxygen, which strips other compounds of hydrogen ... tends to interfere with chemical processes that build complex hydrocarbons from simple hydrocarbons. Is there a name for the sort of atmosphere that is conducive to the generation of complex hydrocarbons from simple ones?"

"Yes," he replied, "the kind of atmosphere to which you are referring is known as a reducing atmosphere. Molecules that can donate hydrogen atoms, or, more precisely, electrons, to other substances tend to dominate that kind of atmosphere.

"Methane and ammonia, the gases used in Miller's experiment, are both considered to be relatively good reducing agents. This means they tend to be involved in chemical reactions involving, to simplify things somewhat, the donation of some of their hydrogen atoms or electrons, which then interact with other hydrocarbon compounds to help make possible, under the appropriate conditions, the formation of even more complex hydrocarbon molecules.

"In one sense, all organic compounds are actually different gradations of reduced forms of carbon. Generally speaking, this is due largely, although not necessarily always, to the presence of hydrogen in such compounds.

"Creating different kinds of reducing atmospheres under experimental conditions, investigators were able to produce a variety of amino acids. Glycine, valine, alanine, proline, glutamic acid and aspartic acid all have been generated through different kinds of electric discharge experiments.

"In another experiment, when sunlight was passed through a solution of paraformaldehyde $(CH_2O)_3$, ammonia (NH_3), and ferric chloride, the amino acids asparagine and serine were produced. On the other hand, when solid ammonium carbonate was exposed to high doses of gamma rays, small quantities of the amino acid, glycine, along with formic acid (HCOOH), resulted.

"In 1961, another scientist, Juan Oró, wondered if amino acids could be generated under laboratory conditions if one used chemical processes that were even simpler than those involved in Miller's earlier experiments. Previous experiments had proven that if one exposed a mixture of hydrogen, nitrogen and carbon monoxide gases to a spark discharge, the reaction would yield hydrogen cyanide (HCN), which is a very reactive intermediate compound.

" Oró combined hydrogen cyanide with ammonia (NH_3) and water (H_2O). This chemical reaction produced a number of different amino acids, just as the Miller experiment had.

"In addition, among the product residues of his experiment, Oró discovered something else. This molecule was a purine ... a nitrogen-containing base of considerable importance.

"The particular purine found by Oró is known as adenine. This molecule is one of two purine bases having a general formula of $C_5H_4N_4$, and three pyrimidine bases, each of which has a general formula of $C_4H_4N_2$. When any of these are combined with either of the pentose sugars, ribose or deoxyribose, together with a phosphate group, then, RNA or DNA is produced.

"Adenine is also one of the components of adenosine triphosphate. This latter molecule is one of the fundamental energy-providing compounds in most organisms.

"In addition to adenine, a number of other useful products could be produced by means of reactions involving hydrogen cyanide. These products included a variety of intermediate precursor molecules that constitute steps on the way to purine or pyrimidine formation, and the products of the reactions included, as well, a number of pyrimidine base molecules that are found in the nucleic acids of some, but not all, living organisms.

"Subsequent experiments demonstrated the possibility of generating, through natural chemical processes, the other nucleic acid bases ... namely, uracil, cytosine, guanine and thymine, which are found in the vast majority of organisms on Earth. These reactions also started with hydrogen cyanide (HCN), but they required, as well, the presence of two other simple carbon compounds: cyanogen (C_2N_2) and cyanoacetylene (HC_3N), which are believed to have been present on the prebiotic Earth.

"Still other experiments were able to demonstrate that the pentose sugar, ribose, an important component of RNA, could be produced quite easily. This chemical process merely involved a series of spontaneous reactions involving molecules of formaldehyde CH_2O.

"Over and over again, scientists were showing, experimentally, the possibility of starting with simple compounds and combining them to produce complex hydrocarbons. More importantly, these products were not just arbitrary molecules, but, rather, they were fundamental building blocks of compounds, such as proteins and nucleic acids, that are crucial to the life process.

"Researchers felt their laboratory experiments were recreating the conditions of prebiotic Earth and demonstrating that chemical reactions important to the origins of life would occur spontaneously. The whole process was relatively simple and straightforward.

"Initially, for example, atmospheric gases, such as methane and ammonia, would react together to generate a variety of simple hydrocarbons, like hydrogen cyanide and molecules known as aldehydes, which are compounds that contain a CHO group ... such as formaldehyde (CH_2O). Next, the products of the first round of reactions ... namely, aldehydes, hydrogen cyanide and ammonia ... would enter into a second round of chemical interactions that would result in such intermediary products as amino nitriles. These products, in turn, would react with the water of the ocean to produce ammonia and amino acids, like glycine.

"People such as Sidney Fox were able to discover, experimentally, alternative methods for the prebiotic generation of various kinds of amino acids - methods that were different from the ones outlined by Miller and Oró. When Fox heated urea [$CO(NH_2)_2$] and malic acid ($C_4H_6O_5$) at temperatures of 150 degrees Celsius, he was able to obtain aspartic acid.

"Fox also was able to construct chains of amino acids through a process of thermal co-polymerization or cooking. He referred to these chains of amino acids as 'proteinoids' because they had certain structural similarities to the proteins found in living organisms.

"The recipe for thermal co-polymerization of amino acids is fairly simple. One starts with some quantity of a given amino acid, such as glutamic acid.

"One places this quantity of amino acids in an oil bath and heats it at 170 degrees Celsius for an hour. When the timer goes off after an hour, one blends in a finely ground mixture of other kinds of amino acids.

"One heats this new mixture for an additional three hours at the same temperature as before. In addition, one heats it in an atmosphere of carbon dioxide.

"When the mixture has cooked for the requisite period of time, one allows it to cool under controlled conditions. When it is ready, one can examine the residue of this process and find polymerized or chemically linked sequences of amino acids of varying length and composition.

"Many of the proteinoid polymer chains consisted of up to 100 amino acids. The nature of the bonds linking the amino acids varied in character, but some peptide linkages, the kind that occur in proteins in living organisms, were observed among these bonds.

"The thermal co-polymerization process is capable of providing yields, by weight, of up to fifteen percent of the total mixture. These portions are considered by evolutionary biologists to be quite ample yields, although most of the rest of us might feel them to be too small to share with friends for a late-night snack.

"There are variations on the foregoing recipe. Glutamine, another of the amino acids occurring in living organisms, is substituted for glutamic acid. Phosphoric acid is also added.

"In addition, one skips the step of pre-heating prior to the adding of other ground-up amino acids. Everything else stays, more or less, the same, yielding roughly similar results as before.

"One can play around with parameters such as the temperature and time, at which and for which, respectively, the mixture is cooked. One also can alter the ratios of the reactants and/or phosphoric acid to be used in the process.

"Experiments focusing on the manipulation of these variables have permitted proteinoids with different kinds of character to be produced. For example, one can increase the percentage of neutral and basic amino acids that were incorporated into the polymerized chain.

"In 1977, a scientist by the name of Usher demonstrated that when one used relatively low temperatures, one could generate phosphodiester bonds between the phosphate and ribose sugar portions of nucleic acids. This is an important step in generating fully functional DNA and RNA molecules.

"In 1978 Juan Oró showed, experimentally, that if one heated fatty acids, an important building block of lipids, and, then, dried them in the presence of phosphate and glycerol, one could synthesize simple phospholipids. Phospholipids are fundamental to the formation of cell membranes in most living organisms.

"Stanley Miller has synthesized a compound under prebiotic conditions that is known as pantetheine. This molecule has been observed to link amino acids in some organisms.

"Many of the compounds produced in these kinds of experiment are quite soluble in water. Researchers have hypothesized that these molecules, at one point or another, probably would have gone into solution in the ocean, and, later, they would have become part of more concentrated solutions when washed, by winds and tides, into the margins of marine lagoons, tidal pools and other intertidal regions, from which water was being evaporated.

"This process of enhanced concentrations through evaporation is thought to be important by many researchers since, quite frequently, the presence of water seems to inhibit the process of polymerization or chaining of, say, molecules. Sidney Fox, along with other scientists, has found, for example, that in order to bring about the polymerization of amino acids, the conditions within the experimental apparatus should be anhydrous ... that is, done in the absence of water.

"Similarly, experimenters have discovered that ribonucleotides will not form oligomers or small chains of up to ten units of nucleic acids unless done by means of anhydrous heating. Furthermore, such heating must occur in the presence of both: a nucleotide triphosphate and cyanamide (CH_2N_2), a condensing agent."

"Dr. Yardley," the prosecuting attorney intervened, "what is a condensing agent?"

"Condensation," answered Professor Yardley, "involves a rearrangement of atoms in order to produce a molecule of greater complexity, density or weight. Condensing agents assist this process.

"Some scientists have hypothesized that exposed mineral, lava or sand surfaces, where temperatures might have reached 100 degrees Celsius, could have served as crucibles on which films of organic compounds, that washed in from the ocean, might have formed covalent bonds. This would have taken place through condensation reactions.

"Furthermore, these researchers theorize that once more complex hydrocarbons formed, some of these molecules might have migrated, through one natural process or another, downward a few centimeters below the surface. Such a micro-environment would have helped to protect the newly-formed compounds from degradation reactions driven by light and heat."

Pausing for a moment, Dr. Yardley finished the remainder of the water in his glass. He replaced the glass on the table in a way that suggested he was thinking about something else.

| Origin of Life |

When he had settled in his seat, his lips were pursed. Finally, he spoke again.

"Even if one were to suppose," he added, "that some of the starting ingredients cited in the previous experiments were not produced in abundance through chemical reactions on prebiotic Earth, one should remember that exogenous or extraterrestrial sources might have helped supplement the normal, earthly complement of these compounds. Water, hydrogen cyanide, ammonia, cyanoacetylene, and formaldehyde ... all of which I mentioned earlier, are found in interstellar dust clouds and might have found their way into meteors, comets or dust particles and, then, subsequently, been transported to Earth.

"Some scientists, in fact, have estimated that in the first 700 million years of the Earth's existence as a planet, the Earth is likely to have passed through 4-5 interstellar clouds, taking roughly 600,000 years to complete each such passage. For each year of passage, this would have resulted in, approximately, between 1-10 million kilograms of material being added to the Earth.

"This is thought to be one or two orders of magnitude, or powers often, less than what has come through interplanetary dust particles. Moreover, like these latter dust particles, only a small percentage of this total mass would be in the form of simple carbon or hydrocarbon molecules. Nonetheless, even a limited percentage of such astronomical figures still would constitute a substantial amount of carbonaceous material available to the prebiotic environment."

"Although," said the lawyer for the prosecution, "I'm quite certain, Professor Yardley, you could provide the jurors with a great deal more information on laboratory experiments that are intended to simulate the conditions on prebiotic Earth, I would like to shift gears slightly. Earlier in your testimony, you had alluded to the importance of having some degree of understanding of the systems of energy that, in a prebiotic environment, would have driven many of the chemical reactions you just have been describing.

"Would you please tell the court a little about this facet of evolutionary thought? Once again, Professor, and I apologize for being a one-note-Norman on this matter, to whatever extent possible, try to strike a balance between avoiding both oversimplification and too much technical complexity."

Dr. Yardley sighed slightly and, then, took a deep breath. He looked briefly at the table by the witness stand, noticed that the pitcher had not much water in it, and made a few motions to Mr. Mayfield indicating he would like the jug to be refilled.

As one of the officers of the court went about the business of getting more water, Professor Yardley started to speak. "There are," he began, "five or six energy possibilities that are likely to have been available to prebiotic Earth for the purposes of bringing about certain kinds of chemical evolution.

"The first possibility requires no external input of energy. These involve physio-chemical forces, such as hydrogen bonds, which, very likely, played a significant role in helping certain molecules in the prebiotic environment to organize or self-assemble into more complex, and biologically relevant, packages.

"For over forty years, thanks to the monumental work of, among others, Watson and Crick, scientists have known that the purine, nucleic base adenine in DNA and RNA pairs spontaneously with the pyrimidine, nucleic base uracil in RNA or the pyrimidine, nucleic base thymine in DNA. Similarly, the purine, nucleic base guanine pairs, in spontaneous fashion, with the pyrimidine, nucleic base in both DNA and RNA.

"These pairings are known as Watson-Crick bonds and are a specific example of hydrogen bonding. The complementary pairs of nucleic bases that are strung along two strands of DNA or RNA are brought together in stable configurations by these bonds and, in the process, help lend the double helical structure to the joining of these strands with which most of us are familiar from school and the media.

"Hydrogen bonds occur as a result of the positive and negative, or dipolar, characteristics that arise in compounds containing hydrogen, oxygen and nitrogen atoms arranged in the right kind of geometrical configuration. More specifically, nitrogen and oxygen are both relatively electronegative in nature, whereas hydrogen tends to be electropositive in character.

"This means oxygen and nitrogen are inclined, under certain circumstances, to draw toward their nuclei a few of the electrons of geometrically well-placed, neighboring hydrogen atoms or molecules. As a result, the affected hydrogen atoms of these neighboring molecules become electropositive and, therefore, have a tendency to establish bonds with other neighboring atoms or molecules that offer electrochemically compatible opportunities.

"These hydrogen bonds bring a certain amount of stability to the manner in which, under certain circumstances, atoms and molecules arrange or organize themselves. Consequently, they are thermodynamically favored arrangements because of their tendency to help stabilize the way energy is distributed in a molecular configuration.

"Hydrogen bonds are characteristic of what are referred to as polar molecules. The polar aspect of these molecules is rooted, as indicated previously, in the process of creating electrochemically-charged dipolar, or positive and negative, regions.

"Polar molecules, such as water and ribonucleic acids, have very different physical and chemical properties from non-polar molecules that do not possess such dipolar regions. Many hydrocarbons that do not contain nitrogen and/or oxygen tend to be non-polar in nature.

"The bottom line on all of this is that hydrogen bonding, of which Watson-Crick pairing in complementary bases of DNA and RNA is an extremely important example, is an instance of a spontaneous, thermodynamically favored generation of greater complexity. A chemical reaction is said to be spontaneous if it can take place without requiring any additional energy.

"The reason a reaction can take place without the need of additional energy is because the energy available to the system has a natural tendency to redistribute itself until no further redistribution of that energy is capable of occurring in a spontaneous fashion. This redistribution process leads to a stable configuration of energy distribution that is why a reaction is said to be thermodynamically favored since, under most circumstances, the thermodynamic nature of chemical reactions is to spontaneously follow whatever pathways are available that will lead to such stability.

"Spontaneous reactions yield energy. In other words, if one measures the potential energy of the final, stabilized state of this kind of reaction, one will find less energy than was present at the beginning of the reaction.

"One of the reasons why the final state of spontaneous reactions is stable is because not all of the energy that is being released remains in a chemically useable form. Some of the released energy is in the form of heat that is unavailable ... that is, it cannot be harnessed to run the reaction in a reverse direction, back to the original, initial state prior to the reaction's commencement.

"The term 'free energy' is often used to refer to the form of energy in a given chemical system that is available to be redistributed, if possible, in a way that allows the system to find, if not already realized, its most stable configuration of energy. This configuration is that point at which the available free energy reaches, through the spontaneous activity characteristic of the system in question, its lowest level consistent with such stability.

"As I indicated previously, in the process of yielding or releasing energy during the time required for a spontaneous reaction to run 'downhill' to its stable state, there is a portion of the released energy that is transformed to a form of energy, namely heat, other than free energy. Entropy is a measure of the amount of energy that has been converted from its free energy form to its non-free energy form.

"Spontaneous reactions always result in a decline of free energy. In other words, the total amount of free energy of the products of a chemical reaction always will be less than the total free energy of the initial reactants of the reaction.

"Consequently, in the process of spontaneously seeking out a stable state of redistributed energy ... that is, a state of lowest possible free energy ... free energy is lost. The entropy, the amount of energy in a non-free form, tends to increase.

"Spontaneous chemical reactions in which energy is released to the environment are known as 'exergonic' reactions. Chemical reactions in which energy needs to be acquired from the environment are known as 'endergonic' reactions.

"One can use the released energy of spontaneous, 'downhill', exergonic reactions to drive 'uphill', non-spontaneous, endergonic reactions. This is referred to as a 'coupled reaction'.

"Non-free forms of energy are generated during both the downhill and the uphill portions of these coupled reactions. Consequently, the total amount of entropy will be increased during the process.

"As long as one has downhill reactions to sponge off, then, uphill reactions are possible. However, in order to keep a sequence of coupled reactions going, one becomes engaged in a constant process of borrowing from Paul to pay Peter who has borrowed from Mary in order to pay George, and so on.

"Non-spontaneous reactions always are in need of arranging a loan of energy from the spontaneous energy generators of the world in order to be able to activate the free energy potential of the non-spontaneous system. When there are no downhill reactions available from which an uphill system can borrow, things come to a sort of dynamic halt known as equilibrium in which its uphill, non-spontaneous character does not change, despite the fact activity still is going on within the system.

"There is a minimum amount of free energy that has to be borrowed by, or introduced into, an uphill, non-spontaneous system in order to bring about a chemical reaction. This minimum amount of energy is known as the free energy of activation or the activation energy.

"One of the major issues of evolutionary theory is to provide plausible accounts of how spontaneous, downhill generations of energy were coupled with non-spontaneous, uphill systems of molecules to generate arrangements of hydrocarbons of increasing complexity. Spontaneous chemical reactions that are thermodynamically favored will take one only so far.

"Therefore, while phenomena such as hydrogen bonding and Watson-Crick pairing are important ways of introducing additional organization into a system without having to borrow additional energy, much more is needed to be able to account for the gradual transition, or evolution, from simple hydrocarbons to the emergence of living systems. Many, if not most, of the chemical reactions that are needed to account for how life arose from a prebiotic environment are of the uphill, non-spontaneous variety rather than the downhill, spontaneous kind, and this means, as suggested earlier, the need to find coupling mechanisms of one sort or another.

"There are a fair number of coupling candidates that would have been readily available under prebiotic conditions. I'll list the candidates first, and, then I'll explore a few of these possibilities.

"First, although not necessarily the most important, are electrical discharges. In a prebiotic environment, these would be manifested through lightning.

"A second candidate would be ultraviolet radiation. Various molecules are capable of absorbing different dimensions of the ultraviolet portion of the spectrum of electromagnetic radiation. When a molecule absorbs ultraviolet light of the right wavelength, the energy of the light can be utilized to help drive certain kinds of chemical reactions involving such a molecule.

"A third possibility for a source of energy capable of driving some non-spontaneous, uphill reactions would be ionizing radiation. Gamma radiation, together with so-called cosmic rays, would be examples of this kind of candidate.

Prebiotic heat would be a fourth coupling candidate. For instance, a surface that had been heated to high temperatures ... either by sunlight, or by a nearby volcano, or by a hydrothermal vent ... such a heated surface might have provided an environment that helped bring about condensation reactions and the forging of various kinds of covalent bonds among molecules lying about on that surface.

"Another possibility involves the energy associated with shock waves. Such waves, for instance, accompany lightning discharges but are distinct from the electrical energy of those discharges.

"In addition, shock waves occur when meteors traverse the Earth's atmosphere. Such waves also are generated when there is an airburst of, say, a carbonaceous chondrite in our atmosphere.

"Tremendous amounts of energy are released under these circumstances. This could be coupled with, and utilized by, various uphill systems.

"There is a further possibility that is not really a source of energy but that would have an important impact on whether or not the minimum energy of activation was achieved in a, heretofore, non-spontaneous, endergonic set of molecules. This additional candidate concerns the process of catalysis.

"A catalyst is capable of helping reactions to proceed by, among other things, helping to lower the normal, minimal level of energy that usually needs to be imported in order to activate a given chemical reaction. A wide variety of non-protein, non-enzymatic mechanisms -- ranging from clays, to metal ions, to RNA ... have been proposed as possible catalytic agents in a prebiotic environment.

"Since, previously, I already have given something of a taste for what is possible, experimentally, with the electrical discharges of Miller's experiment and the anhydrous, heat driven experiments of Fox, I would like to touch on a few of the other possibilities. Once again, this treatment won't be exhaustive, but it will provide members of the jury with a framework of sorts through which to understand this aspect of the evolutionary model.

"There have been a number of laboratory experiments that explored certain aspects of the phenomenon of shock waves. For instance, the heat, in the vicinity of 3000 degrees Kelvin or more that is generated by rapidly expanding gases in shock wave tubes has been used to produce such hydrocarbons as hydrogen cyanide and amino acids in different kinds of gas mixtures with reducing properties.

"A few researchers have hypothesized that organic compounds might have been synthesized in the atmosphere when meteors passed by and, in the process, created conditions similar to the shock heating experiments in the laboratory. After being synthesized, these compounds would have found their way, through one means or another, to the ocean.

"Once in the ocean, one of three things is likely to have occurred. The newly synthesized molecules would have reacted further with molecules in the ocean; or, these molecules would have been carried to tidal pools and other intertidal zones where they would become concentrated and readied for further reactions when the water in these pools and zones evaporated; or, some combination of the first two possibilities.

"Some scientists have calculated that meteors with a mass between: 10^{-14} to 10^2 grams, enter the Earth's atmosphere with sufficient frequency to deliver about 1.6×10^7 kilograms of mass to the Earth each year. If one were to assume these meteors traveled with a velocity of 15 kilometers per second, the meteors collectively generate about 1.8×10^{15} joules of energy per year, which is equivalent to many megatons of explosives.

"Ah ... Professor, before you continue," Mr. Mayfield interrupted, "could you explain what a joule is."

"A joule," Dr. Yardley explained, "is a unit of work or energy equivalent to the work that is done, or the heat generated, in one second, by an electric current of one ampere against a resistance of one ohm and ... " Stopping, Professor Yardley smiled sheepishly and raised his eyebrows somewhat. "Sorry," he said, "I don't think my answer is quite what you were looking for, Mr. Mayfield."

After thinking about the matter for a few seconds, the professor informed the lawyer: "The easiest, maybe most recognizable, thing to say" he offered, "is this. A watt of energy is equivalent to 1 joule per second. However, one should keep in mind that, strictly speaking, a watt is a measure of power, whereas a joule is a measure of energy. Power deals with the rate at which energy is expended."

Dr. Yardley looked at the prosecution lawyer with a more hopeful expression, seeking, apparently, acceptance for his new approach. When Mr. Mayfield motioned his head and made a face, both of which seemed to suggest: Why don't we move along before things get worse, Dr. Yardley returned to his testimony concerning the energy created by atmospheric shock waves.

"100 percent of the kinetic energy of meteorites of the previously indicated size is lost to the atmosphere. Researchers maintain that some fraction of this energy is converted into the generation of atmospheric s hock waves. Estimates of the fraction of the energy being converted in this manner run from 30% downward.

"Working along similar lines, researchers have made calculations for the amount of energy that is converted to shock waves for other kinds of phenomena. For example, the airbursts of carbonaceous chondrites with a radius that is less than, or equal to, 300 meters, is believed to generate about 1.5×10^{14} joules of energy per year, which is the equivalent of a huge number of high explosives.

"When a meteorite does not airburst and strikes the ground, if the meteorite is sufficiently big in size, it will generate a post-impact vapor plume. Some researchers have calculated that such post-impact vapor plumes could generate as much energy as 6×10^{17} joules per year in the form of shock waves that, once again, would be the equivalent of many megatons of high explosives.

"In addition to the energy being converted into shock waves capable of synthesizing certain organic molecules, researchers have estimated that a small percentage of the carbon in the meteorite will be incorporated into organic compounds when the meteorite vaporizes upon impact. This percentage is considered to be about 4%, which would have yielded approximately 4.6×10^6 kilograms of organic materials per year on prebiotic Earth.

"Most of this incorporated carbon shows up in the form of carbon dioxide and carbon monoxide. However, several percent of the carbon are incorporated into various kinds of hydrocarbons, and there is a still smaller percentage being converted into such compounds as hydrogen cyanide, as well as aldehydes, like formaldehyde.

"If one adds all of these different kinds of energy and mass values together, one can begin to develop a thermochemical model of shock synthesis under both reducing and relatively neutral atmospheric conditions. Scientists have discovered that the efficiency with which organic compounds can be synthesized through shock waves is very dependent on the compositional character of the atmosphere in which the shock wave occurs.

"For example, in a reducing atmosphere of methane, nitrogen and water vapor, for each joule of energy generated by shock waves, one can produce approximately $10^{17.5}$ molecules of hydrogen cyanide. Simultaneously, lesser amounts of simple hydrocarbons like C_2H_2, C_2H_4, and carbon soot also will be produced.

"After all the calculations are done, this works out to be a yield of 1.2×10^{-8} kilograms of organic material is generated for each joule of shock-created energy in a reducing atmosphere. In a neutral atmosphere, on the other hand, consisting of, for instance, carbon dioxide, nitrogen and water vapor, a yield of 2.5×10^{-16} kilograms of hydrogen cyanide is produced for each joule of energy, but yields of formaldehyde (H_2CO) remain roughly equivalent to what occurs in a reducing atmosphere.

"Similar calculations have been carried out in relation to both lightning and coronal discharges in the atmosphere. For example, in a reducing atmosphere, lightning is estimated to have been likely to generate 3×10^9 kilograms per year of organic material from the 1×10^{18} joules per year of energy created.

"In a neutral atmosphere, lightning is calculated to have been likely to produce 3×10^7 kilograms of organic material per year from the same amount of energy yields. However, as the atmospheric ratio of hydrogen gas relative to carbon dioxide drops from, say, 2 down to 0.1, the yield of hydrogen cyanide, formaldehyde and amino acids drops by a factor of several magnitudes or powers of ten.

"If one combines all the different ways of using energy that would have been available on prebiotic Earth to generate organic materials, scientists estimate that about 10^{11} kilograms of organic material would have been produced each year in a reducing atmosphere. However, in a relatively neutral atmosphere, consisting of mostly carbon dioxide and about 10% hydrogen gas, approximately 10^9 kilograms of organic materials would have been produced each year, but this yield will fall considerably as the relative percentage of hydrogen gas drops.

"In the light of these calculations, evolutionary scientists have come to the following conclusion. If all the organic materials produced by these various means were fully soluble in oceans comparable in extent and depth to our present oceans, and if these organic materials had a mean lifetime of approximately 10^7 years with respect to thermal degradation in relation to mid-ocean hydrothermal vents, then the steady-state equilibrium of organic materials in prebiotic times would have been about 10^{-6} grams of organic solute for each gram of ocean water in a neutral atmosphere, and approximately 10^{-3} grams of organic solute for each gram of ocean water in a reducing atmosphere.

"Modern researchers in evolutionary theory believe that if the early Archean era atmosphere were strongly reducing in character, the predominant method of generating organic materials might have been through shock waves. Lightning would have been considerably less predominant in its effects in this regard, and the roles of ionizing radiation and radioactive disintegration would have been quite negligible."

"So," said Mr. Mayfield, "if someone wanted to put all of this information into perspective in a relatively simple manner, what would be the bottom line?"

"I guess" replied Professor after a few seconds hesitation, "one should return to the scenario I outlined earlier. Moreover, for the sake of simplicity, let's concentrate on just one of the hydrocarbons, namely hydrogen cyanide, that is likely to have been produced by one, or more, of the energy sources about which I have been talking.

"First, energy from shock waves or lightning or ultraviolet radiation is coupled with atmospheric gases such as methane (CH_4), ammonia (NH_3), and hydrogen (H_2), all of which serve as reducing agents, giving up hydrogen atoms or electrons to other substances. This coupling leads to the production of hydrogen cyanide.

"Secondly, the HCN or hydrogen cyanide that is formed becomes dissolved in water vapor in the atmosphere. Eventually, this becomes precipitation or rain that falls into the ocean.

"Thirdly, once in the ocean, the hydrogen cyanide would oligomerize or gather together in small quantities here and there. These oligomers of HCN would then undergo hydrolysis in the ocean.

"Hydrolysis is a process in which water interacts with a substance and tends to separate out the atoms of a substance such as hydrogen cyanide (HCN) by hydrating them, that is, surrounding them with water molecules. Furthermore, since water is a polar molecule involving, as previously indicated, dipolar regions of electronegative and electropositive charge, the polar character of water combines with the atoms that are being separated out through the process of hydrolysis to recombine to form different kinds of molecules.

"For instance, just to give you some idea of what is being said here, suppose one had a one-liter solution of one-tenth molar concentration of hydrogen cyanide and left it for a year. As a result of hydrolysis, after one year, one would find quite tiny, but detectable, amounts of the purine nucleic base adenine as well as larger, but still very small, quantities of the amino acid glycine.

"If we project such liter-size processes into the context of the trillions and trillions of liters of the oceans of the world, and if we left things for millions of years rather than one year, we are very likely to discover substantial amounts of a wide variety of complex hydrocarbons, many of which probably will be of fundamental importance to issues concerning the origins of life."

"Dr. Yardley, in the context of the present discussion, what relevance would the process known as a 'Strecker synthesis' have?" asked Mr. Mayfield.

"In synthetic, organic chemistry," responded the professor, "a Strecker synthesis generally involves bringing about the hydrolysis of, say, an amino nitrile in the presence of a strong acid. An amino nitrile joins together some kind of amino group or radical with a cyanogen or compound containing the group CN.

"Many researchers have accepted the pH value of the early, prebiotic ocean to be around 8, plus or minus 1. This means that the ancient ocean was considered to be either slightly basic, if it had a pH of 8-9, or relatively neutral, if its pH was around 7.

"Under such conditions, Strecker synthesis, which usually is done in the presence of a strong acid, would require a long time to hydrolyze organic compounds in the early, prebiotic ocean. Some researchers have set this figure at around 10,000 years.

"However, relative to tens and hundreds of millions of years, 10,000 years is really just a drop in the ocean so to speak. This kind of synthesis would have had the opportunity to run to completion many times over during the course of the Archean era.

"I should note, Mr. Mayfield, that although the Strecker synthesis process is considered by evolutionary theorists to be an adequate means of producing amino acids in the ancient oceans, some sort of additional mechanism of concentration and condensation would be required to produce, say, the purine, nucleic base, adenine. This is where processes such as evaporation, freezing and dehydration, along with hot, anhydrous conditions, which are believed to have been present in certain intertidal zones, would play important roles in chemical evolution on early earth."

"At this point, Dr. Yardley," requested the lawyer, "would you say a little about current thinking in relation to the nature and possible origins of membranes? I believe such a discussion will bring us a little closer to providing the jurors with a proper, introductory overview of evolutionary theory, by means of which they will be able to reach an informed judgment on the matter before the court."

"I suppose," Professor Yardley mused, "that molecules known as amphiphiles are as good as place as any with which to begin talking about the origins of membranes. Amphiphiles have sort of an aura of split personality about them.

"One part of this kind of molecule has hydrophilic properties and, as a result, is inclined to enter into interactions with water. The other part of the molecule entails hydrophobic characteristics and, therefore, tends to avoid, whenever possible, interacting with water.

"When amphiphiles are immersed in an aqueous environment, the hydrophobic aspects of the molecule curl up into small spheres known as vesicles. These tiny spheres form a protected space within which, given the right conditions and chemical reactants, various chemical processes could take place.

"The hydrophilic-portion of amphiphiles -- that is, the parts of the molecule that have an affinity for water – surrounds the hydrophobic aspects of the molecule. Not only do these hydrophilic portions represent an additional layer of separation between water and the interior, hydrophobic aspects of the amphiphile molecule, the water- loving components of the molecule also are free to enter into reactions with water.

"As such, amphiphile molecules possess some of the basic features of biological membranes. More specifically, the membranes of living organisms tend to be bilayered or have two membranes that are separated from one another by a relatively short distance, and the whole bilayered structure surrounds the interior of the cell.

"To be sure, the layered arrangement in amphiphile molecules is not quite the same as the sandwich structure of biological membranes. Most notably, there is no separate, distinct region between the hydrophilic and hydrophobic components of the amphiphile molecule, as there is in true, biological membranes.

"Nonetheless, in both true membranes as well as amphiphile molecules, one does have a double- layered arrangement surrounding an interior space or spaces within which chemical reactions could take place. Furthermore, both true membranes as well as amphiphile molecules consist of hydrophilic and hydrophobic aspects.

"Consequently, amphiphile molecules could be considered to constitute a rather crude facsimile or early precursor, of later, more complexly evolved, biological membranes. Interestingly enough, in this respect, some researchers maintain that exogenous organic materials -- that is, organic materials from sources such as meteorites and interplanetary dust particles -- might be quite rich, perhaps even preferentially so, in amphiphilic vesicles or spheres.

"Lipids, which are one of the main components in biological membranes, come in different varieties. As far as biological membranes are concerned, some of the more important lipids are composed of, among other things, a hydrophobic hydrocarbon component linked to a hydrophilic phosphate group, together with certain alcohols and/or bases.

"Lipids do not form polymers or chains of monomer units as, say, amino acids and nucleic acids do. This is because lipids are stabilized through non-polar physical forces instead of the covalent chemical bonding that characterizes polymerized compounds.

"These non-polar physical forces are essentially thermodynamic in nature. Non-polar hydrocarbons, such as oil, do not enter into solution when placed in water ... water being a polar molecule.

"Hydrocarbons have a tendency to disrupt, at least in part, the array of hydrogen bonding present in water. The most stable thermodynamic arrangement ... that is, the arrangement in which all of the molecules of a system have achieved their lowest chemical potential for reactivity ... is one in which hydrocarbon molecules aggregate into a separate phase form, such as droplets, away from water molecules.

"This process of phase separation between non-polar hydrocarbon molecules and polar water molecules is known as the hydrophobic effect. This effect serves as a significant force helping to stabilize various kinds of macro molecular systems, including membranes, in biological organisms.

"The hydrophobic effect does not involve any chemical transformations. It only reflects the natural preference, or self-organizational drive, of molecules to arrange themselves in ways that distribute the energy of the molecular system in the least chemically reactive, and, therefore, most stable state.

"Some evolutionary scientists have suggested that the hydrophobic, hydrocarbon portion of lipids might have been synthesized or formed by a Fischer-Tropsch-like reaction. This process starts with carbon monoxide and hydrogen that are placed under pressure, ranging from one to fifty atmospheres, as well as heated to temperatures that might vary from 180 to 300 degrees Centigrade.

"Usually, this reaction is done in conjunction with a catalyst of some sort. Many catalytic possibilities exist, but, quite frequently, the ones that are used are either nickel or iron supported by a layer of silica.

"Phospholipids, which are one of the fundamental building blocks of biological membranes, come in several forms. They are polar molecules in which the phosphate group has a negative charge, the alcohol group has a positive charge, and the complex hydrocarbon tail is hydrophobic in nature.

"More importantly, phospholipids, once formed, have been observed to assemble, spontaneously, into stable lipid bilayers and vesicles within an environment of water because of the aforementioned thermodynamic forces that are at work. The hydrophilic components of the molecule form the portions of the bilayer that will be in close proximity to water molecules, whereas the hydrophobic portions of the molecule form the aspects of the bilayer that will be phase separated from water molecules.

"Some scientists have approached the issue of the first, primitive cellular prototype from a different direction than that of amphiphilic molecules composed of both hydrophobic and hydrophilic components. These researchers have focused on certain kinds of proteinoid micro spheres that have been observed to form under certain experimental conditions.

"Once again, this sort of protocell structure would form a phase separation between the outer, aqueous environment and the inner regions of the micro sphere formed by the proteinoids. These inner regions could serve as a location for various kinds of chemical reaction to take place under conditions that are, to some extent, protected and stable.

"All membranes of living organisms consist of a combination of phospholipids and proteins. Therefore, if one were to combine the idea of proteinoid micro spheres and amphiphilic molecules, one would be getting quite close, in some respects, to the structural character of modern biological membranes.

"A cell is really a microenvironment bounded by a membrane. The phospholipid portion of the membrane constitutes a permeability barrier that helps stabilize and protect the microenvironment of the cell's interior.

"However, the down side of a permeability barrier is that it can keep out various kinds of molecules that might be necessary to chemical reactions going on in the interior regions of the bounded micro-environment. In biological cells, this problem is solved by a variety of proteins, referred to as transmembrane proteins, which extend from one membrane layer to the other membrane layer of the bilayered structure.

"These transmembrane proteins might serve different functions. Some of them provide channel ways, linking the external aqueous world with the internal bounded microenvironment.

"Some of these membrane proteins might function as carriers, or active transports, for certain kinds of molecules. Still other forms of these membrane proteins might form part of an ion pump system that brings various ions into the cell or gets rid of such ions de pending on circumstances and needs.

"When, as evolutionary biologists believe to be the case, proteinoids, at some point, became incorporated into self-assembling, phospholipid membrane structures, a major step would have been taken toward the first protocell. Various experiments of nature might have ensued then, exploring different arrangements and kinds of proteinoids in the membrane, some of which were naturally selected because of their ability to serve, in some minimum fashion, as channels, or carriers or parts of an ion pump system.

"Researchers feel those proteinoids would have been favored that had particular kinds of primary structure. More specifically, the sequence of amino acids constituting the primary structure of the protein should be such that, under the influence of purely thermodynamic, self-organizing forces, the tertiary folding pattern brought about by these thermodynamic forces would need to have arranged hydrophilic and hydrophobic aspects of the proteinoid in a certain manner.

"On the one hand, hydrophilic portions of the transmembrane proteinoid would need to be at the opposite ends of the membrane where they would be exposed to water molecules surrounding the cell as well as within the cell. On the other hand, those portions of the proteinoid structure that were hydrophobic should be folded away in the region between the two bilayers ... a region that consists of hydrophobic lipid molecules.

"Prior to the appearance of such phospholipid-proteinoid micro spheres, there might have been transitional structures. Liposomes, for example, are small vesicles composed of fluid, lipid bilayers.

"Liposomes have the capacity for reversible breakage. In other words, under various conditions, they can break open and, then, spontaneously reseal.

"Thus, when liposome vesicles are agitated in an aqueous environment, they will break open at various points and, afterward, reseal.

This process of breaking and resealing enables the liposome to capture any solutes that might happen to be in the environment.

"Similarly, when liposomes are dried, they often form multi-layered structures. Solutes can become trapped within these structures. When the dried liposome becomes re-hydrated, the trapped solutes become sealed within the microenvironment of the liposome's interior.

"The property of being able to break and reseal could serve another function beyond providing a mechanism for admitting different kinds of solute materials into the liposome's interior region. Growth, division and multiplication, of a sort, also could be associated with this capacity to break and reseal.

"If one were to add some of the potential properties of a liposome to those of phospholipids and proteinoids, then the possibilities become even more intriguing. Such an amalgamation of properties is coming much closer to what we would recognize as a cell-like structure or protocell."

"I believe," indicated Mr. Mayfield, "we are almost to the end of our conceptual journey, Professor Yardley." The prosecuting lawyer went to his table and was handed some papers by his colleague.

As Mr. Mayfield slowly returned to the area of the witness stand, he was busy going through the papers. Apparently, he was either looking for something or briefly reviewing the material prior to launching into the next phase of direct examination.

When he was near the witness stand, he stopped and studied the papers for a few more seconds. When he had finished, he asked: "Professor Yardley, what is the so-called Central Dogma of molecular biology?"

"Essentially," Dr. Yardley replied, "it says that DNA makes RNA that makes proteins. In living organisms, the available evidence is overwhelmingly in support of this principle."

Pursuing the issue, Mr. Mayfield inquired: "Does this principle raise any problems in relation to accounts concerning the origins of life?"

"Yes, it does," the professor responded. "In everyday terms, it leads to the question: Which came first, the chicken or the egg?"

"Could you elaborate a little, Professor Yardley?" requested the prosecuting lawyer.

"Briefly stated," the professor summarized, "if DNA is necessary to make, first, RNA and, then, proteins, but the synthesis of DNA polymers depends on the presence of catalytic or enzymatic proteins, then, how can one start with DNA that is dependent on the very molecule that it is supposed to make? On the other hand, if proteins depend on the existence of DNA and RNA molecules, then how can proteins come into being prior to that on which they depend?

"If we have DNA reprise the role of the egg to protein's stirring rendition of the chicken, we, once again, are faced with an ancient paradox. In the present case, the problem becomes: which came first, the protein or the DNA?"

"Is there any plausible way out of this dilemma, Dr. Yardley?" asked the prosecuting attorney.

"Until relatively recently, this paradox constituted a major stumbling block to providing an overall plausible explanation for how life originated from prebiotic beginnings through purely naturally processes. The situation vis-à-vis this paradox began to change around 1983.

"In that year, two researchers, Sidney Altman and Thomas Cech, quite independently of one another, made a breakthrough. They discovered what has come to be known as a 'ribozyme'.

"A ribozyme is a polymerized or chained sequence of molecules that is drawn from RNA and exhibits some of the properties of a protein enzyme or catalyst. In the case of the Altman-Cech findings, the RNA sequence that had been discovered was able to cut and join pre - existing strands of RNA.

"This ability to cut into a given sequence of RNA and, then, to splice such sequences together is of considerable importance. Broadly speaking, not only do such capacities allow for the possibility of building longer sequences of RNA, but, as well, cutting and splicing constitute tools that could play fundamental roles in processes of both replication and the rearranging of ribonucleic acid sequences to generate, or experiment with, alternative genetic characteristics.

"Most importantly, ribozymes do not presuppose anything else to accomplish these functions. In other words, the chicken/egg paradox evaporates since an RNA sequence that is capable of acting as an enzymatic molecule in relation to other RNA molecules, depends on neither proteins nor DNA in order to come into being. RNA molecules are serving as both hereditary blueprints as well as catalytic agents for the generation and development of such blueprints.

"RNA molecules have a further advantage, at least in relation to DNA molecules. The ribonucleotides in RNA ... that is, the bonded triads of ribose sugar, phosphate and nucleic base that are chained together to create sequences of RNA ... such ribonucleotides are more easily synthesized than are the deoxyribonucleotides of DNA.

"On the other hand, deoxyribonucleic acids are more stable than are ribonucleic acids. Consequently, whereas the easier path of synthesis would have conferred an evolutionary advantage on RNA molecules over DNA, the property of greater stability would have conferred, later on, an evolutionary advantage of DNA molecules over RNA.

"Many theorists cite this dimension of greater stability as probably one of the primary factors that led to a gradual evolutionary transition from RNA-based protocells or life to DNA-based protocells or life. At some point, DNA displaced RNA from the latter's role as keeper of the genetic memory.

"During the 1960s, some twenty years before the discovery of the first ribozyme, three scientists, Francis Crick, Carl Woese and Leslie Orgel, all working independently of one another, had each suggested that RNA might have had evolutionary priority over both DNA and proteins. Today, the original proposal of these three scientists has evolved, through the contributions of a variety of theorists and researchers, to become a theory known as 'the RNA world'.

"In the RNA world, as one might anticipate, RNA plays a central role. RNA, as the carrier of genetic information, as well as the agent responsible for catalyzing reactions, becomes responsible for generating all of the steps considered necessary to produce the first precursor of life capable of self-replication and evolutionary change.

"One can lend support to the idea of the RNA world theory with a number of recent findings. For example, consider the work of Harry Noller Jr..

"He was doing research on ribosomes that frequently are called the protein factories of a cell. Ribosomes consist of, on the one hand, ribosomal RNA, which differs in certain ways from non-ribosomal RNA, and, on the other hand they contain various kinds of protein. Both of these components are joined together to form what are known as ribonucleoprotein subunits.

"Each ribosome is assembled from two such subunits. Each of these subunits is slightly different in size and kind from the other.

"Furthermore, different kinds of ribosomal subunits can be found in prokaryotic, or non-nucleated organisms, and in eukaryotic, or nucleated organisms. However, just to complicate matters, some of the former, prokaryotic kinds of ribosomal units also can be found within certain eukaryotic intracellular organelles, or membraned centers, such as the energy-related factories known as mitochondria and chloroplast.

"In general terms, ribosomes travel along the length of various strands of messenger RNA. Messenger RNA is a single-stranded transcription of the triplet nucleic bases that are carried by DNA molecules. These triplet, nucleic base sequences, or codons, constitute the letters, so to speak, designating the specific word from the dictionary of twenty amino acids that is being called for by means of the messenger RNA.

"As a ribosome travels along the strand of messenger RNA, the ribosome helps forge a linkage, known as a peptide bond, between amino acids. The ribosome does this by taking the amino acid called for by one triplet nucleic base sequence of messenger RNA and connecting the indicated amino acid with another amino acid that is being called for by the subsequent RNA triplet nucleic base sequence of the same strand of messenger RNA.

"The ribosome accomplishes its task of fashioning polymers or chains of amino acids ... that is, proteins ... with the assistance of a further kind of RNA, known as transfer RNA. This form of RNA consists of between 70-80 nucleotides that are specially modified or adapted to be able to interact with the construction area formed by both the ribosome and the strand of messenger RNA.

"One portion of the transfer RNA carries the amino acid being called for. Another section, known as the anticodon, links up with the appropriate codon section of the messenger RNA, and the final sequence of the transfer RNA links up with the ribosome.

"Thus, transfer RNA delivers the required amino acid to the active site of the interaction between the ribosome and messenger RNA. This tri-partite co-operative effort continues, using a succession of different transfer RNA molecules, until the fully formed protein, which is being specified by the collective set of triplet codons of messenger RNA, has been completed.

"Harry Noller, the scientist I mentioned earlier doing research into the nature and functioning of these ribosomes about which I have been talking, discovered something of considerable importance to the RNA world theory. He found evidence suggesting that ribosomal RNA appears to play a major catalytic role leading to the formation of peptide bonds between amino acids being delivered by transfer RNA to the site of interaction between messenger RNA and the ribosome.

"The proteins present in ribosomes also have a catalytic role to play. Yet, this role appears to be limited to one of enhancing the degree of efficiency of the process already set in motion by the ribosomal component of the subunit.

"This molecule, consisting of ribosomal RNA and protein, is known as ribonuclease-P, and it is considered to be a true enzyme. Not only does it accelerate the rate of the formation of peptide bonds significantly over what would occur in the absence of such a molecule, but, as well, the molecule survives the chemical reaction and is capable of repeating the process with other transfer RNA molecules.

"On the one hand, the self-splicing ribozyme mentioned earlier is not considered a true enzyme. Although that molecule does have an enzyme-like function that involves capacities for cutting and splicing, nonetheless, at the end of the chemical reaction, the molecule does not get restored to its original, pre-reaction form.

"On the other hand, scientists, like Gerald Joyce, have been able to take this research a few steps further. Through a variety of procedures, he was able to generate ribozymes ... that is, RNA with catalytic properties that could cleave a number of different kinds of chemical bonds, including the peptide bonds that link amino acids together in biological organisms.

"In 1993, researchers at the Scripps Research Institute in California synthetically created a small sequence of RNA, sometimes referred to as the Scripps molecule, which had some amazing properties. First, the molecule began to make copies of itself within an hour after it had formed.

"Secondly, the copies of this molecule began to make copies of the copies. Finally, these copies began to evolve and display a variety of chemical properties that had not been anticipated.

"In another development, around 1994, Jack Szostak isolated a relatively short sequence of nucleotides, known as an oligonucleotide, which had catalytic-like properties. This catalyst could join together other, short sequences or oligonucleotides.

"In addition, this same catalytic agent could utilize energy from a triphosphate group in order to underwrite the polymerizing or chaining character of that molecule. This is important because triphosphates play fundamental roles as suppliers of energy for chemical reactions taking place in a living cell.

"Other researchers have proposed alternative routes for, say, the synthesis of RNA oligonucleotides that could be considered complementary to the previous findings. For example, James Ferris discovered that montmorillonite -- a relatively, common clay -- is capable of synthesizing RNA oligonucleotides."

"Dr. Yardley," said the prosecuting lawyer, "I believe we have covered enough information to provide the jurors with a good, though necessarily abbreviated, overview of the evolutionary perspective concerning the origins of life from prebiotic beginnings. If you were to sum up the general thrust of your testimony, what would you say?"

The professor stared off into the space near the ceiling at the back of the court room. After about ten seconds of deliberation, he stated: "If one runs through the available evidence in support of evolutionary theory, of which my testimony is but a very small sampling ... if one considers all the cosmological, geological, meteorological, hydrological and chemical data, then, I believe there is only one way to make consistent sense of the existing evidence.

"Biological organisms arose gradually, as the result of a series of steps, each of which was selected by prevailing circumstances that favored such a step over other possibilities existing at the time of selection. This fortuitous confluence of natural forces required tens of millions, if not hundreds of millions, of years to complete themselves.

"Among other things, this confluence of forces included various kinds of energy interacting with the gases in the atmosphere to generate

| Origin of Life |

simple hydrocarbons. These hydrocarbons subsequently precipitated out into a set of hydrological conditions that, perhaps through a Strecker synthesis process, were conducive to the formation of a sequence of progressively more complex hydrocarbons, such as amino acids, purines and pyrimidines.

"In addition, when these complex hydrocarbons were subjected to further processes of dehydration and condensation in various intertidal zonal regions, then, eventually, a variety of proteins, nucleic acids, lipids, and carbohydrates formed that were incorporated into bounded -- or membraned -- micro-environments from which arose the first protocells capable of self-replication. This capacity, very likely, was as a result of, initially, RNA catalytic activity that, at some point, became transformed into a DNA-based living organism.

"Throughout the sequence of gradual, evolutionary steps leading from the formation of the Earth, to the first cellular system capable of self-replication and genetic experimentation, spontaneous processes of self-organization played important roles. In other words, although chemical kinetics ... the study of the paths and rates of actual reactions ... constitutes an essential part of evolutionary thinking, nevertheless, thermodynamic forces also spontaneously led to arrangements of energy distribution that had important evolutionary ramifications for the forms and functions that different molecules came to have.

"Although there are certain details of the foregoing scenario that are presently eluding our grasp, we -- that is, evolutionary scientists -- believe all of the basic components are, in principle, now present for a rigorous, consistent, and plausible account of the origins of life through purely natural processes. Moreover, scientists and researchers, collectively, are quite confident, despite the fact there might be certain details that currently are missing from our account, that these same details will be forthcoming in the near future by virtue of the sort of scientific discoveries that are being made every day around the world."

"Thank you, very much, Dr. Yardley," said the prosecuting attorney, "for your illuminating, expert testimony." As the lawyer walked back to his table, he said: "Your witness, counselor."

The attorney for the defense was about to rise, when the judge said:

"Mr. Tappin, we are approaching -- if not encroaching on -- the dinner hour. Before you start your cross-examination, I think we will adjourn for meals.

"The jury is instructed not to discuss these proceedings either among themselves or with anyone else. Court will be in recess until 7:30 p.m. " With that pronouncement, she banged her gavel.

Ah, Sweet Mysteries of Life

Judge Arnsberger entered the courtroom, and everyone had risen in concert with the command to do so that was given by one of the court officers. Again, in obedience to a directive, we all sat down.

"Mr. Tappin" stated the judge, "you may begin your cross-examination. Dr. Yardley, please remember, you still are testifying under oath."

Picking up a note pad from the table in front of him as he arose, Mr. Tappin approached the witness stand. Smiling at the professor, the defense lawyer said: "Dr. Yardley, I would like to commend you on an excellent presentation during direct examination."

The professor angled and dipped his head slightly in acknowledgment of the compliment. The smile on his lips was a tentative one, and the look in his eyes was wary in character.

The two looked like a cobra and a mongoose ready to do battle. Which was which, was a toss-up.

Beginning the conceptual competition, Mr. Tappin briefly referred to the note pad he was carrying and stated: "In your discussion concerning meteorite impacts of the early Earth, Professor, you indicated that the scientific models dealing with what was happening on Earth, and when, were based on various studies conducted in relation to the lunar cratering data acquired through the Apollo space program. Is this correct, Dr. Yardley?" the lawyer asked.

"Yes, that's right," the professor answered.

"To the best of your knowledge," inquired Mr. Tappin, "what is the oldest time frame for which a radiometric date has been fixed in relation to the lunar samples?"

"That would be the Apollo 16 and 17 uplands data," Dr. Yardley responded. "The radiometric dating process has established a time frame of between 3.85 and 4.25 billion years ago for the lunar samples taken from the craters in the areas of the two, aforementioned Apollo expeditions."

"Do the samples from the uplands represent the most heavily cratered areas of the lunar surface?" Mr. Tappin asked.

"No, they don't," the Professor indicated.

"Therefore, Dr. Yardley, am I right in assuming that, at the present time, we don't have any radiometric data from these more heavily cratered areas of the moon?"

"Your assumption is correct," affirmed the professor.

"Then, this would seem to suggest," the lawyer stated, "that we don't know whether the more heavily cratered areas are older or younger than the lunar samples that have been brought back to Earth, or, perhaps, a bit of both ... that is, some craters might be older, and some might be younger."

"Yes, at present, the age or ages of the more heavily cratered areas of the moon only can be estimated," the Professor acknowledged. "More precise dates must come from radiometric testing of samples from those areas."

"How would one go about estimating the age of areas of the lunar surface for which we have no direct data?" Mr. Tappin inquired.

"Well, this is really not my area of specialization," pointed out Dr. Yardley, "but, I suppose, a lot would depend on one's choice of decay rates and how one fitted this to the available lunar cratering data."

"Dr. Yardley, would the choice of decay rates substantially affect one's conclusions, both with respect to amounts and times, in relation to the models of extraterrestrial bombardment of early Earth?"

"Whether or not one's conclusions would be affected substantially, depends on what one means by the word 'substantially'," the professor replied. "In general, however, the use of different methodological or radiometric starting points obviously will have some kind of impact on one's conclusions."

"If I understand you, Professor," Mr. Tappin said, "the choice of decay rates with respect to lunar cratering data could increase or decrease estimates of such variables as: how many meteorites, what size and when such meteorites collided with the Earth. Is this, essentially, the case?"

"In broad terms, yes," Dr. Yardley confirmed. "As I indicated in my earlier testimony, the model concerning the influx of meteorites into the Earth's atmosphere is largely a stochastic or probabilistic one.

"Consequently, a range of values is possible," indicated the professor. "The ones I have given to Mr. Mayfield are best-estimate projections based on carefully worked out models of probability that are believed to have governed what transpired on early Earth as far as meteorite activity is concerned."

"Dr. Yardley, in your direct testimony, I believe you stated many evolutionary researchers are of the opinion that much of the heavy meteorite bombardment of early Earth probably began to taper off somewhere between 4.44 billion and 3.8 billion years ago. Is this true?"

"Yes," the professor affirmed.

"You also testified, did you not Dr. Yardley, that many scientists contend an extremely large meteoric impact occurred on Earth approximately 65 million years ago off the Yucatan peninsula, and there is evidence to indicate this collision might have destroyed most of the species in existence on Earth at the time?"

"I gave such testimony, yes," admitted the professor.

"Was the Yucatan crater the result of a statistical anomaly?" asked the defense lawyer. "In other words, can we assume that between, say, 3.8 billion years ago and 65 million years ago, there were probably few, if any, large-sized meteoric impacts on Earth?"

"Such an assumption would be a reasonable one," the professor said.

"What makes the assumption reasonable, Dr. Yardley?" inquired the lawyer.

"Well, for one thing," Dr. Yardley answered, "the very fact life continues to exist, and, on the basis of paleontological data, has existed for over 3.5 billion years, indicates there cannot have been too many large-sized meteorite collisions with Earth. If there had been, we probably wouldn't be having this conversation."

"In your opinion, Professor, would living organisms have a better chance of surviving such a catastrophic event than various prebiotic arrangements of complex hydrocarbons?" Mr. Tappin asked.

At this point, Mr. Mayfield jumped up and firmly stated: "Objection, Your Honor. The question is highly hypothetical and speculative."

"Mr. Tappin" probed Judge Arnsberger, "do you care to respond to the objection?"

"Yes, Your Honor, I do," replied the defense lawyer. "On the basis of both direct testimony, as well as on the basis of evidence derived from cross-examination to this point, the nature of science has been shown to involve, among other things, the use of assumptions, hypothesis, conjecture, probability, projections, estimates, interpolations and extrapolations. Therefore, I fail to see on what plausible grounds the prosecution could object to the defense's desire to explore certain hypothetical and speculative issues concerning the origin-of-life problem from a scientific perspective."

"Mr. Tappin has a point, Mr. Mayfield," the judge indicated. "I'm inclined to cut him some slack on this line of questioning provided the attorney for the defense doesn't roam too far astray.

"Objection overruled. The witness should answer the question," she stated.

Turning his attention from the judge to the lawyer for the defense, Dr. Yardley replied: "In my opinion, the answer to your question would depend on quite a few variables. For example, one factor would concern whether the size of the meteor impact was sufficiently large to vaporize the ocean, or merely big enough to boil, to the point of evaporation, the 200-meter layer beneath the ocean's surface known as the photic zone."

"Excuse me, Professor," interrupted the defense lawyer, "what is the photic zone?"

"The 200-meter photic zone represents the depth to which light penetrates with sufficient energy to be able to sustain photosynthetic autotrophs. Photosynthetic autotrophs are organisms that synthesize their organic requirements by using sunlight as a source of energy to convert inorganic materials, such as carbon dioxide, to molecular forms capable of being used by the organism to sustain itself."

"Thank you," said Mr. Tappin, "please continue."

"The first kind of impact mentioned previously ... that is, one capable of vaporizing the ocean, would involve, roughly speaking, about 5×10^{27} joules of energy. This amount of energy would be delivered by an object that was around 440 kilometers in diameter and/or had a mass of 1.3×10^{20} kilograms, traveling at approximately 17 kilometers per second.

"The second kind of impact ... that is, one capable of boiling away the photic zone, would require about 4×10^{26} joules of energy. The object would have a mass of approximately 1.1×10^{19} kilograms and a diameter of about 190 kilometers.

"The Chicxulub, Yucatan crater, by way of comparison, is calculated to have been created from an object that is some 300 kilometers in diameter. Thus, it is intermediate in size between meteorites capable of evaporating the ocean and meteorites able to boil away the 200-meter photic zone near the ocean's surface.

"If the size of the impact were of the ocean-evaporating kind, then, neither living organisms nor various complex arrangements of hydrocarbons would have been likely to survive to any appreciable degree. To understand why this is so, one needs a few facts about the nature of the collision being discussed.

"With an impact of this magnitude, roughly a quarter of the energy arising from the collision would have been directed toward vaporizing the water of the ocean. Another quarter of the impact energy would have been radiated upward toward the atmosphere, and the remaining fifty percent of the energy would be buried in the vicinity of the impact.

"The heat generated at the point of impact would be sufficiently great to melt, if not vaporize, most of the crustal material ejected from the crater being formed by the force of the collision. The temperature of these materials probably would reach around 2000 degrees Kelvin or 1727 degrees Celsius.

"Furthermore, the heat released through these melting and vaporizing materials would have been radiated in at least two directions. There is a thermal wave of some 2000 degrees Kelvin that would have been generated upward toward the atmosphere, as well a thermal wave that would have been radiated downward.

"The rock vapor that radiated upward would have surrounded the globe for a period of time, raising the atmospheric temperature considerably. By the time the rock vapor had rained out, so to speak, from the atmosphere, half of the ocean would have existed in the form of a hot steam that would have added about 140 times of our present sea level pressure to the atmosphere.

"A short while after the rain out of the rock vapor, which would take several months, the uppermost portions of the steam atmosphere would have cooled enough to generate a relatively thick, moist zone capable of convectively reflecting substantial amounts of heat back to Earth. A number of researchers believe this would have led to the runaway greenhouse threshold, or beyond, at which time the rest of the ocean would boil away.

"There are a number of factors that could affect the character of the foregoing sequence of events. The amount of carbon dioxide in the atmosphere would be one consideration, especially given that the manner in which CO_2 is distributed among earth, atmosphere and the ocean is quite complex, with different greenhouse and temperature scenarios following from different modalities of distribution.

"In addition, the amount of cloud cover -- as well as whether the cloud cover was at higher or lower altitudes -- could affect the amount of infrared radiation that is absorbed and radiated back to Earth. On the other hand, cloud cover also could affect the amount of sunlight that might be reflected away from the Earth.

"Eventually, depending on the actual atmospheric temperature, pressure, and so on, the water content of the atmosphere would begin to precipitate out and fall back to Earth and, in this way, reform the ocean. This period of cooling and ocean re-formation would probably take between 2,000 and 3,000 years to be completed.

"The impact of a meteorite sufficiently large to boil away the 200- meter photic zone of the ocean also would have catastrophic results, although, obviously, not quite as pronounced as those that I have just described. For one thing, after an impact of the lesser kind now being addressed, the atmospheric disturbances and restoration of the ocean to relatively 'normal' conditions would take merely 300 years, rather than 2-3000 years as previously indicated for the larger kind of impact.

"If the nature of an existing ecosystem is such that it is dependent, ultimately, on photosynthetic autotrophs, then, the sterilization of the photic-zone would wipe out the ecosystem. In other words, when the bottom link of the food chain in a given ecosystem disappears, then all of the heterotrophs higher up the chain that depend on that link also will disappear."

Professor Yardley noticed the expression on the face of the defense attorney. The professor seemed to reflect for a second on what he had just said.

Upon, apparently, intuiting the question about to be asked, he started to speak again. "Heterotrophs," he added, are organisms that depend on other life forms, usually photosynthetic or chemosynthetic autotrophs, to provide them with the organic materials that can be used to derive energy by which to synthesize their organic needs."

Mr. Tappin smiled in acknowledgement of Dr. Yardley's correct intuition. The defense attorney gave a slight motion of his hand indicating for the witness to proceed.

"However," pointed out the professor, "not all life forms live within the photic zone, and not all life forms necessarily are dependent on photosynthetic autotrophs in order to survive. There are chemosynthetic autotrophs, involving a few species of bacteria, which derive their source of energy for organic synthesis completely independently of light energy.

"These organisms accomplish this by means of the oxidation of various reduced inorganic compounds. For instance, some of these chemosynthetic autotrophs, like the colorless sulfur bacteria, have the capacity to generate energy by oxidizing hydrogen sulfide to sulfur, while other organisms, like certain nitrifying bacteria, possess the ability to produce energy through oxidizing ammonia to nitrite.

"If these chemosynthetic autotrophs lived far enough below the photic zone, or lived sufficiently deep beneath the earth's surface, so as not to be affected by an impact large enough to vaporize the photic zone of the ocean, then such organisms might stand a very good chance of surviving this sort of catastrophic event. Similarly, complex, prebiotic hydrocarbons located out of harm's way in the same fashion as these chemoautotrophs also would be likely to survive a collision of this lesser kind.

"Thank you, Dr. Yardley," said the defense lawyer, "I believe you have answered my question quite adequately. Now, let's see if I understand the overall character of this part of your position as stated in direct testimony.

"Earlier, you informed Mr. Mayfield and the court that researchers have concluded, based on lunar radiometric analysis, there were as many as 15-16 meteorite collisions on Earth that were greater than the impact creating the largest crater on the moon and, therefore, might have been sufficiently big to evaporate the oceans of early Earth. You further testified, Dr. Yardley, that researchers contend the last of these ocean-vaporizing events probably took place somewhere between 4.44 billion and

3.8 billion years ago. Is my understanding correct on both of these points, Professor?" inquired Mr. Tappin.

"Yes, it is," Dr. Yardley agreed.

"On the other hand," the lawyer proceeded to say "none of this would preclude any number of lesser collisions capable of sterilizing the photic-zone from having occurred. Presumably, the Yucatan crater serves as indirect evidence for such a statement since it was considerably larger than what is minimally necessary to boil away the photic-zone and, yet, here we are talking about it. Would I, more or less, be correct in asserting this, Dr. Yardley?"

"In general," replied the professor, "I would be prepared to go along with you except I would add one proviso to what you have said."

"Yes, Professor, what would this proviso be?" inquired the lawyer.

"If one had too many impacts capable of sterilizing the photic - zone," suggested the professor, "then, this could prove to be as problematic, in its own way, to the development of life or to the development of prebiotic systems as were impacts of the ocean - vaporizing variety. Such impacts do occur ... as the Chicxulub, Yucatan crater demonstrates ... but we believe the available evidence indicates these kinds of collisions, probably, were relatively rare events after 3.8 billion years ago, the time when the last of the ocean-vaporizing events is believed to have occurred."

"Yet, Dr. Yardley," the defense lawyer said, "the fact of the matter is there really is very little, if any, available evidence to indicate how many impacts there might have been, from, say, 3.8 billion to 3.5 billion years ago, which were capable of boiling away the photic zone. Is this not correct, Professor? Yes or no?"

"You would have ..." Dr. Yardley began to say. The defense attorney interrupted.

"Your Honor, I find the witness' answer non-responsive," Mr. Tappin stated.

"Dr. Yardley," Judge Arnsberger explained, "you must answer the queries of the defense counsel in accordance with the form in which the questions are being asked. In this particular case, your only options are 'yes' or 'no'"

"Thank you, Your Honor," acknowledged Mr. Tappin. "Would you like me to repeat the question, Dr. Yardley" he asked.

Shaking his head in a negative fashion, the professor sighed and said: "Yes."

"So, to restate the matter, Professor," the lawyer paraphrased, "statistically speaking, there might have been: no impacts, or one impact, or a few impacts, or more than a few impacts, of a size sufficient to boil away the photic zone of the ocean during the indicated period between 3.85 billion and 3.5 billion years ago. Is this correct?"

"Yes, that is correct," Dr. Yardley replied.

Flipping the page on his note pad, Mr. Tappin scanned the contents of the page for a few seconds and said: "Professor, in your earlier testimony concerning indirect, isotopic evidence for the existence of life 3.85 billion years ago which has been discovered at the Isua rock formation in Greenland, you mentioned, in passing, certain kinds of methodological contraindications with respect to the previously stated interpretation of that evidence. Would you explain" requested Mr. Tappin, "at this time, a bit more about the nature of these possible counter-indications?"

"As I said earlier," noted Dr. Yardley, "during the fixation of atmospheric carbon dioxide, living organisms tend to discriminate against the Carbon^{13} isotope and prefer its Carbon^{12} counterpart. This is due to the kinetic character of the enzyme responsible for the fixation of carbon in so-called C_3 plants ... that is, plants in which a three-carbon acid is the first product of photosynthesis.

"Consequently, one will find organic sediments exhibiting depleted amounts of Carbon^{13} relative to atmospheric CO_2. On the other hand, inorganic carbonate sediments, such as limestone, will tend to display elevated levels of Carbon^{13} relative to atmospheric CO_2.

"If one encounters a sample that fits the depleted Carbon^{13} profile, such evidence can be interpreted to mean that the profile was produced by a C_3-like plant that has a carbon-fixing enzyme with this tendency. The issue, unfortunately, is not always straightforward.

"This is especially true in cases where the sample is drawn from a rock formation, such as Isua, where the rocks have, at some time, been subjected to temperatures in the range of 450 to 700 degrees Celsius. Such high temperatures might bring about what is referred to as a partial re-equilibrium of any carbon isotopes that are present in the rock formation.

"This partial re-equilibrium of carbon isotopes tends to elevate the Carbon^{13} values for organic samples. At the same time, this process causes a lowering of the Carbon^{13} value for the inorganic carbonate sample.

"When this happens, the results are skewed. Under such circumstances, one might not know if one is dealing with an inorganic carbonate with a lowered Carbon^{13} value, or if one is dealing with an organic material with an elevated Carbon^{13} value.

"Some people have interpreted the Isua carbon isotope evidence to mean that the samples in question were produced by a carbon–fixing enzyme similar in character to the enzyme existing in C_3 plants of today. Other investigators are not so sure if this interpretation is correct."

"What ramifications follow from these different interpretations, Dr. Yardley?" inquired the lawyer for the defense.

"If the first interpretation I mentioned is true -- that is, if the Isua sample is actually organic in origin -- then, evidence would have been established that pushes back the earliest known life form to at least 3.85 billion years ago, several hundred million years, and change, prior to our previous oldest, fossil evidence drawn from the Warrawoona Group in Western Australia. If, on the other hand, the Isua sample turned out to be an inorganic carbonate with thermally skewed low Carbon^{13} values, giving a false positive for organic matter, then, the oldest known evidence for the existence of life would stand at around 3.55 billion years ago, give or take thirty million years, or so."

"If," hypothesized Mr. Tappin, "the organic interpretation of the Isua isotope evidence is correct, then presumably, this would suggest an upper boundary had been established for ocean-vaporizing meteorite impacts. In other words, given the catastrophic character of this kind of collision, as outlined by you previously, then one would be hard-pressed to account for the continued existence of photosynthetic life forms like the proposed Isua organism. Would you agree with this, Dr. Yardley?"

"Yes, I concur," the professor indicated.

"On the other hand," offered the lawyer, "depending on circumstances, the location, the hardiness, and the luck of our hypothesized Isua organism, this photosynthetic autotroph might or might not survive an impact capable of vaporizing the photic zone. Is this correct?"

"Yes, I think so," stated Dr. Yardley.

"Now, Professor," continued the lawyer, "this approximate date of 3.85 billion years ago puts us at the upper, or later, limit of the period between 4.2 billion and 3.8 billion years ago that you cited earlier as the time during which the last of the 15-16 ocean-vaporizing meteorite collisions with earth is projected to have occurred. If one were to claim the final ocean-vaporizing impact were to have occurred some 4.2 billion years ago, then one has, approximately, 425 million years to play with in order to account for the origin-of-life. Is this right, Dr. Yardley?"

"Right," replied the professor.

"However," remarked Mr. Tappin, "on the one hand, there is no compelling evidence to suggest one would be justified in adopting the earlier 4.2 billion-year bench mark as one's starting point. On the other hand, there is some evidence ... namely, projected photic-zone vaporizing and ocean-vaporizing meteorite collision like the one near the Yucatan Peninsula some 65 million years ago ... suggesting the parameter of 4.2 billion years might be a tad premature. Do you feel my characterization of the situation, Dr. Yardley, is unfair?"

"Not really," admitted the professor. "The starting point for origin - of-life scenarios has considerable theoretical and empirical looseness to it."

"If," Mr. Tappin conjectured, "scientists suddenly were to discover evidence indicating the incorrectness of the organic interpretation of the Isua sample, then, in your opinion Dr. Yardley, would the arbitrary nature of this starting point issue change much?"

"Yes and no," the professor responded.

"Would you please elaborate," requested Mr. Tappin.

"The fixing of a time frame that establishes a non-catastrophic period of time having conditions conducive to a prebiotic account of the origin-of-life always will have an element of arbitrariness about it. Nevertheless, using the later 3.55 billion-year Warrawoona date as the time when life initially had become firmly established is friendlier to evolutionary models than is the Isua date of 3.85 billion years ago.

"The later, Warrawoona dating of life fits in more comfortably with the available data than does the earlier, Isua dating. By this, I mean the earlier dating of life has more problems to overcome in a shorter period of time than does the later dating of life.

"Among other things, the earlier, Isua dating of life is overlapping with the meteorite impact data, which we have discussed, much more than is the later, Warrawoona dating of life. There are more likely to have been both ocean-vaporizing and photic-zone vaporizing impacts associated with the earlier, Isua-dating than with the later, Warrawoona dating of life."

"Still, Professor Yardley, wouldn't you agree," inquired the defense attorney, "that one of the bottom lines in all of this is the following? In the light of the meteorite impact data, do we really have any non-arbitrary way to theoretically determine the amount of time with which we have to play around, so to speak, as far as providing a plausible evolutionary account of the origin-of-life is concerned?

"In other words, are we not merely guessing in relation to the basic question? Do we have any empirical means of pinning down how much historical or Archean time we actually have to work with in order to provide an account of the transition from prebiotic conditions to the first protocell or full-fledged organism that is plausible?

"Isn't one as justified in saying there was only 4,000 years, or less, say, between the last catastrophic meteorite impact and the laying down of the physical evidence, whether direct or indirect, for the first appearance of life on Earth, as one is claiming there was some 425 million years between these two points in history? Aren't evolutionary scientists arbitrarily selecting the latter time interval, during which life allegedly arose simply because it proves to be less embarrassing and problematic for their theory than the 4,000-year scenario would be?"

"I believe," responded the professor, "there is a difference between: making educated, empirically based conjectures about the origin -of-life and creating myths concerning those origins. I maintain there is a difference between, on the one hand, making conjectures with respect to which one can seek out evidence both for or against, and, on the other hand, developing systems of beliefs that are removed from empirical data as well as from rigorous demonstration."

"Dr. Yardley," interjected the defense lawyer, "one can agree entirely with what you just have said, but you haven't addressed the essential thrust of my previous line of questioning. Let me restate the issue in another manner.

| Origin of Life |

"Fact one: in your testimony, Professor Yardley, you indicated researchers have maintained there probably were 15-16 meteorite collisions with the Earth occurring sometime after 4.3 billion years ago. Furthermore, these collisions were projected to possess more force than the ones causing the largest lunar crater Imbrium.

"Fact two: the magnitude of these events would be sufficient, at the higher level, to vaporize the ocean, or, at the lower level, to vaporize the photic zone.

"Fact three: these collisions were believed to have occurred somewhere between 4.3 and 3.8 billion years ago.

"Fact four: these events were stochastically distributed across a 500 million-year interval.

"Fact five: the first indirect, potential evidence for the existence of life is dated around 3.85 billion years ago.

"Fact six: the first, direct fossil evidence for the existence of life is dated from about 3.55 billion years ago.

"Fact seven: an event intermediate between a collision that would have vaporized the ocean and one that would have vaporized the photic zone occurred approximately 65 million years ago.

"My questions to you Dr. Yardley are these: One, given the foregoing facts, when precisely, during the interval between, say, 4.3 billion years ago and 3.55 billion years ago, did the 15-16 projected collisions with Earth occur?

"Two, given the foregoing facts, is one justified in treating the event that took place 65 million years ago, as part of the stochastic distribution of the original 15-16 events?"

Dr. Yardley looked at Mr. Tappin, apparently considering the questions. The professor started to speak and, then, stopped.

Finally, he said: "There really is no way, at the present time, to answer your first question with any precision. As far as the second question is concerned, I'm not sure the Yucatan crater should be considered as part of the original stochastic distribution profile.

"I suppose, nonetheless, a case might be made by some individuals to include, on justifiable grounds, the Yucatan event in the original stochastic distribution. The object that collided with Earth some 65 million years ago might well have been a remnant of the original debris that had been

| Origin of Life |

bombarding the Earth during the Archean era and on which the projected 15-16 collisions is based."

"Would you agree, then, Dr. Yardley," inquired the defense lawyer, "that, on the basis of the available evidence, someone who claimed the last ocean-vaporizing collision took place 3.73 billion years ago would be as justified in her or his claim as the person who claimed the last ocean-vaporizing collision took place 3.54 billion years ago?"

"Yes and no," replied the professor. When Dr. Yardley realized Mr. Tappin was waiting for the answer to be expanded on, the professor said: "I agree, reluctantly, with your basic point about the unknown nature of the historical time that was actually available to be able to go from prebiotic conditions to biological organisms through natural processes.

"On the other hand," the professor added, "if the Isua sample does have organic origins, then, the person who claimed the last ocean - vaporizing event took place 3.54 billion years ago is somewhat in conflict with the facts because of the evidence for the existence of life at both 3.85 billion years ago, as well as 3.55 billion years ago. Seemingly, a continuity of some sort has been established through the two kinds of dated evidence for the existence of life at Isua and Warrawoona."

"Isn't it conceivable," asked Mr. Tappin, "that life might have originated more than once? After all, Professor, in your direct testimony you spoke about the possibility of protocells and organisms existing in the early Archean era that were not part of the lineage that is linked, in any way, with the last common ancestor of all modern forms of life. Were you not suggesting during your testimony that life could have arisen, in various forms, more than once?"

"Yes," Dr. Yardley acknowledged, "I was suggesting this. However, the fossil evidence discovered at the 3.55 billion-year old Warrawoona Group contains the imprints of eleven different kinds of microorganisms. One would be asking a lot to suppose this much diversity could arise so quickly after an ocean-vaporizing event of the sort you have hypothesized."

"I agree with you," confirmed Mr. Tappin. "Such a scenario might be stretching things to the point of snapping, but this is not my problem, Professor, it is yours.

| Origin of Life |

"You are the one who says he has a plausible account of, or explanation for, the origin-of-life from prebiotic beginnings. The viability of that claim is what is being probed through this cross-examination."

Without pausing, Mr. Tappin pressed on. "Dr. Yardley," he asked, "are you familiar with the so-called 'faint young sun paradox'?" "Yes, I am," responded the professor.

"Would you explain to the court the nature of this paradox?" Mr. Tappin requested.

"On the basis of various calculations performed by astronomers, many scientists accept as likely that 4 billion years ago, the sun actually was some 25-30 percent dimmer than today. If this is so, then a possible paradox emerges.

"More specifically, considered in terms of the current atmospheric conditions of the world, if the sun were 25-30 percent dimmer than is presently the case, then, the upper 300 meters of the ocean would freeze, along with rivers, lakes and inland seas. In addition, under these circumstances, the ice sheet covering the Earth would reflect much of the rest of the sun's incoming light, thereby preventing any thawing from taking place.

"Evidence, on the other hand, derived from a variety of sedimentary rocks indicates liquid water was in existence around 3.8 billion years ago. Furthermore, direct fossil evidence demonstrates the existence of biological organisms as early as 3.55 billion years ago.

"The paradox is as follows. How could liquid water and biological organisms exist in environmental conditions that should have been frozen due to the presence of a faint young sun?"

"Is it not possible," inquired Mr. Tappin, "that various combinations of hydrothermal vents, volcanic islands, and so on, in different parts of the Earth, could have generated a set of relatively localized conditions capable of, over time, producing both sedimentary rocks as well as sustaining life forms?"

Professor Yardley shrugged his shoulders. His face had an expression that seemed to be a blend both of skepticism as well as a considering of possibilities in relation to the defense attorney's suggestion.

The professor's head bobbed back and forth slightly, and he appeared to be weighing things in his mind. Finally, he said: "Maybe, but researchers have come up with a number of other possibilities."

"Would you outline a few of these possibilities?" requested the defense attorney.

"Since astronomers calculate the early sun probably would not have overcome its faintness until around 2.5 billion years ago," Dr. Yardley began, "the challenge is to devise ways capable of permitting the Earth to compensate for the sun's relative dimness during the Archean era. The ways that have been devised concern conjectures about the compositional character of the paleoatmosphere -- that is, the Earth's early atmosphere.

"For example, during the 1970s, there were several attempts to resolve the faint early sun paradox. The first proposal focused on methane and ammonia, while a second suggestion concerned carbon dioxide.

"Ammonia and methane both absorb, and, therefore, trap, certain portions of the infrared spectrum that is being produced by the Earth as the planet is heated by solar radiation. The absorbed infrared energy heats up the atmosphere, and the atmosphere, in turn, begins to radiate infrared wave lengths, some of which return to the Earth's surface in the form of what many people have referred to as the 'greenhouse effect'.

"If there were enough methane and ammonia in the atmosphere, then considerable amounts of infrared energy would be absorbed and, eventually, radiated back to the Earth. In fact, some researchers believe this process might have been able to generate and radiate sufficient heat back to the Earth's surface to compensate for the faint early sun.

"There are, however, several problems with the methane/ammonia compensation hypothesis. To begin with, both methane and ammonia are susceptible, in varying degrees, to photolytic dissociation, or breakdown, as a result of the effect of ultraviolet radiation.

"Moreover, both methane and ammonia tend to enter into reactions with the hydroxyl radical [OH] that arises as a result of the photolysis -- or breakdown by ultraviolet radiation -- of H_2O. While some of these hydroxyl radicals would combine with the hydrogen gas coming from volcanic emissions, enough free hydroxyl radicals still might have been available for chemical reaction with a great deal of methane and ammonia, and, consequently, removed these molecules from the atmosphere.

"In addition, ammonia is quite soluble in water. Therefore, NH₃ tends to be lost from the atmosphere through rainout.

"There have been some studies indicating that the presence of protective buffers, such as water vapor in the case of methane, and hydrogen sulfide in the case of ammonia, can affect the rates and extent of photo destruction of methane and ammonia. Furthermore, another study suggested the photolysis of methane could produce several hydrocarbons, such as hydrogen gas and methylene (CH_2), which are efficient absorbers of infrared radiation.

"Despite this sort of data, the overall effect of photolysis, chemical reactions and rainout, likely would have resulted in the removal of most of the methane and ammonia molecules that might have been present, at some point, in the Archean atmosphere. Therefore, an atmosphere composed largely of methane and ammonia would not have had a very long lifetime unless there was some continuous source of production for these molecules.

"Today's atmosphere consists of a mixing ratio of about 1 part per billion of ammonia as well as 1.6 parts per million of methane. The presence of these molecules in our atmosphere is entirely the result of biogenic production.

"Once the Earth had differentiated, through the formation of the magnetic core, and, in the process, removed much of the Earth's iron from the surface, there would have been no chemical mechanism on prebiotic Earth, of which I am aware, capable of producing, on a continuous basis, either ammonia or methane.

Thus, these molecules wouldn't be able to solve the faint early sun paradox.

"An alternative theory to the methane/ammonia hypothesis, which also arose during the 1970s, focused on the possible role of carbon dioxide as a means of compensating for the dimness of the faint early sun. Carbon dioxide, like methane and ammonia, is capable of absorbing infrared energy being radiated from the surface of the Earth and, as such, is a greenhouse gas.

"For reasons closely related to the elimination of methane from the theoretical picture, carbon dioxide became a strong candidate for providing a means of compensating for the coolness of the faint early sun. More specifically, when methane is oxidized by the presence of [OH] radicals created through the ultraviolet photolysis of water vapor, carbon dioxide is a product.

"Thus, the oxidation of much of the methane in the early Archean era is considered by many researchers to be a good candidate for helping to generate a considerable amount of carbon dioxide. To this, one can add the substantial portions of volcanic emissions that consist of carbon dioxide.

"In modern times, currently active volcanoes have been estimated to release some 4×10^{10} kilograms of carbon per year. Most of this is in the form of CO_2.

"One reasonably could assume that the amount of carbon dioxide released through volcanic activity during the Archean era was, undoubtedly, far greater than is the case today. Nevertheless, almost any estimates one came up with in this regard would be both speculative and arbitrary to a large extent.

"Furthermore, how much of this out-gassed carbon dioxide would have remained in the atmosphere during the Archean era depends on the amount of this material that would have entered into solution with the ocean, as well as on the amount of carbon dioxide that became incorporated into inorganic carbonate formations such as limestone. Unfortunately, knowing the amounts of carbon dioxide that are in any given form ... gas, solid or liquid ... at any given time is fairly difficult to pin down with any precision in the best of times, let alone some 4 billion years ago.

"Estimates of the amount of carbon dioxide in the atmosphere during the Archean era vary over a wide set of possibilities. Some people believe the amount of carbon dioxide in the prebiotic atmosphere rapidly decreased during the Archean era and remained at relatively low levels thereafter. Other researchers maintain the amount of carbon dioxide at the beginning of the Archean era was high and continued to remain relatively high for some time.

"Among those theorists who contend the amount of carbon dioxide in the atmosphere was fairly substantial, there are again differences in projected amounts. There are researchers who indicate there might have been as much as 100 bars, or 100 standard atmospheres, worth of carbon dioxide gas in the Archean atmosphere. Others suggest the amount of carbon dioxide in the ancient atmosphere might have been between 10 and 20 bars or standard atmospheres.

"A 100-bar atmosphere of carbon dioxide would result in surface temperatures of about 230 degrees Celsius. With a more modest 10-to-20-bars of atmosphere, the Earth's surface temperature is likely to have ranged between, say, 85 and 110 degrees Celsius.

"Both of these scenarios would create surface conditions capable of compensating for the coolness of the faint early sun, thereby eliminating the paradox created by the existence of sedimentary rocks and fossil evidence. Furthermore, even the 100-bar carbon dioxide atmosphere would not necessarily generate temperatures that automatically lead to a runaway greenhouse effect in which all of the surface waters would boil away and be present in the form of clouds or steam.

"The saturation water vapor pressure under such circumstances would be about 30 bars, or so. Consequently, when one adds this to the existing 100 bars of pressure of carbon dioxide, the temperature would have to be raised another 100 degrees before the ocean would start to boil under that kind of pressure."

"Didn't you indicate, Dr. Yardley, that the impact of a meteorite somewhat larger than one capable of vaporizing the photic zone of the ocean would generate a transient rise in temperature of 100 degrees?" asked Mr. Tappin.

"Yes," the professor confirmed.

"So," the defense lawyer suggested, "in the context of a 100-bar carbon dioxide atmosphere, the impact of a meteorite smaller than the one that created the Yucatan crater might be capable of triggering a runaway greenhouse effect?"

"Possibly," stated the professor. "The actual outcome might depend on a lot of different factors."

"All right, Dr. Yardley, let's see if I have this right," Mr. Tappin said.

"Firstly, the early sun is thought to have generated 25-30 percent less luminosity than the sun of today. Under current circumstances, a sun this dim would have resulted in the freezing, among other things, of the oceans to a depth of 300 meters.

"Secondly, the 'big freeze' could have been avoided by an atmosphere with the right kind of compositional character. In other words, the faint early sun paradox could be avoided if the Earth's atmosphere contained enough greenhouse gases to be able to, first, absorb from the

Earth, and, then, radiate back to the planet, sufficient levels of infrared energy to compensate for the 25-30 percent dimmer luminosity of the early sun.

"Thirdly, there are, in broad terms, two competing theories concerning the compositional make-up of the Archean era atmosphere. One theory champions methane and ammonia as the greenhouse gases of choice, while the alternative theory advocates carbon dioxide.

"Fourthly, in neither theory do we know, except in very broad terms, what the precise character of the composition, temperature or pressure of the Archean era atmosphere was. On the other hand, in both cases, there would have been enough infrared radiation absorbed and re-emitted by the respective gases of each theory to counter the cooling effects of the faint early sun.

"Finally, there are substantial arguments for, and against, each of the competing theories. Dr. Yardley, does my brief summary capture the gist of the matter vis-à-vis the faint early sun paradox?" inquired Mr. Tappin.

"Yes," acknowledged the professor, "I would say you have captured all of the highlights."

"Is there any preference among researchers between either of the two theories outlined by you, Dr. Yardley?" asked the defense lawyer.

"The early preference," noted the professor "had been for the methane/ammonia hypothesis. Relatively, recently, however, the preference scales have been tipping rather heavily in the direction of the carbon dioxide perspective."

"Does anything rest on these preferences beyond resolving the faint early sun paradox issue?" wondered Mr. Tappin.

"Quite a bit, actually," stated the professor. "The methane/ammonia hypothesis is far more conducive to providing plausible accounts for the evolution of prebiotic systems than is the carbon dioxide hypothesis.

"The methane/ammonia atmosphere constitutes a reducing environment. Due to the way this kind of atmosphere provides an environment that is conducive to chemical reactions believed to be capable of leading to increasingly complex organic molecular forms, a methane/ammonia atmosphere lends fundamental support to the emergence, eventually, of a variety of biologically important complex hydrocarbons such as amino acids, purines, pyrimidines, ribose sugars and so on.

"On the other hand, a carbon dioxide atmosphere is, at best -- and depending on what other molecules are considered to be present in such an atmosphere ... only slightly reducing, and, therefore, much less conducive, and, perhaps, even antagonistic, to the gradual build-up of the increasingly complex molecular forms required by evolutionary theory. In general, the more hydrogen gas there is postulated to be in a carbon dioxide dominated atmosphere, the greater will be, up to a point, the reducing character of that atmosphere. Alternatively, the more the ratio of [H_2] to [CO_2] falls away from 1, the less reducing will such an atmosphere be.

"Many researchers believe nitrogen, not hydrogen, was the most common gas next to carbon dioxide in the Archean era atmosphere. A nitrogen/carbon dioxide dominated atmosphere would have been either neutral or, possibly, according to some researchers, quite reactive with a propensity to breakdown, rather than build up, more complex hydrocarbons such as amino acids.

"As a matter of fact," pointed out Dr. Yardley, "this problematic dimension of a carbon dioxide dominated atmosphere inspired a couple of theorists ... around 1994, I think ... to develop another approach to the faint early sun paradox. In effect, these researchers seemed to feel there was no need to try to find ways of compensating for the cooling effect of a faint early sun.

"The starting point for their theory is to assume the Earth froze as a result of the early sun's 25-30 percent lower luminosity. The freezing would have created a 300-meter thick layer of ice near the ocean's surface.

"According to the architects of this theory, the layer of frozen ice would have served to protect chemical activity going on in the water below the frozen zone. In addition, the cold, but unfrozen, ocean water would have helped to preserve whatever organic molecules were formed since the decomposition of organic molecules is slower at these lower temperatures.

"Furthermore, these theorists allowed for the influx of large meteorites every million years or so. These large-scale impacts would have melted the ice and helped stir things up, so to speak, in a variety of ways involving shock-synthesis of various hydrocarbons, mixing of organic materials, energy distribution and so on."

"How can one be sure," queried Mr. Tappin, "there would have been any ocean at all beneath the frozen zone, or if there were liquid ocean water below such a zone, how would one know how deep the water would be? If the chill caused by the faint early sun was present from the very beginning of the Archean era, then how would this affect the formation of the ocean?"

"The answer to your question," remarked Dr. Yardley "would depend on a lot of different factors. For instance, scientists believe the process of core formation is likely to have raised the overall temperature of the planet to some 1500 degrees Celsius.

"Obviously, things would have to cool down considerably before lasting bodies of water could have begun to form on the surface. Before this point had been reached, there probably would be a time when the water being released into the atmosphere as a by-product of the core formation process would exceed the saturation level for water vapor in the atmosphere.

"The precise character of this saturation level would depend on things such as atmospheric temperature, pressure and composition. Once such a level was exceeded, then, for a time, there probably would have been a rapid precipitation and evaporation cycle in which water would not have collected on the surface, but humidity would have been quite pronounced.

"At some juncture, surface temperature, as well as atmospheric composition, temperature and pressure, along with water formation and precipitation would have collaborated to create conditions conducive to the generation of relatively stable bodies of water. How one factors a faint early sun into this process of ocean formation is rather difficult to say because so many of the variables being considered are uncertain.

"I'm sure a number of computer models concerning the nature of ocean formation in the Archean era have been developed. Depending on starting assumptions, different models likely would designate different depths of water as the point that would have to be reached before a frozen layer starts to form.

"Hydrothermal vents and volcanic activity also would have to be thrown into the mix since they both are capable of affecting water temperature, locally and, possibly, even globally. With each new variable that is added, the model becomes more complicated and, consequently, providing an answer to your question is less and less straightforward."

"Given," began Mr. Tappin, "what you have been saying, Dr. Yardley, in response to my question, is one being unfair to the facts if one were to argue that the manner in which one pieces together those facts is very much dependent on, or driven by, the assumptions one makes concerning the nature of conditions in which one believes those facts are embedded?"

"No," indicated the professor, "someone arguing in the fashion in which you have suggested would not be treating the facts unfairly. Indeed, in science, one constantly should be examining the relationship between established facts and the assumptions surrounding one's use of, or interpretation of, those facts."

"Could one," asked the defense lawyer, "not also say the following? When the facts of a matter have not been established clearly, then, the relation among assumptions, interpretation and 'facts' becomes, potentially, quite problematic?"

"Yes, I would agree with that," Dr. Yardley replied.

"Therefore, in the matter at hand, Professor... namely, the question of whether or not a 300-meter frozen zone would have formed near the surface of the Archean-ocean as a result of the dimness of the faint early sun ... we appear to be faced with a rather problematic situation. This is so, because given, as seems to be the case, that we don't know such things as the composition, temperature and pressure of the Archean - atmosphere; or, the rate of Archean-ocean formation; or, the water vapor saturation levels of the Archean-atmosphere; or, the degree to which hydrothermal vents or volcanic activity are present, and so on; then, in a very real sense, except in extremely broad terms, we don't know the facts of the matter, do we?"

"No, we don't," confirmed Dr. Yardley. "This is one of the reasons theoretical models are constructed.

"Scientists take what is known about the laws of nature, together with whatever data might be available concerning the conditions surrounding a particular problem, such as the present issue of a frozen zone above the Archean era ocean. Next, certain assumptions are made about how natural laws might be manifesting themselves under certain conditions.

"The implications of these assumptions are worked out in the form of a model. Essentially, the model says that if certain assumptions are true, then, under specified conditions, natural laws will generate certain kinds of predictable activity in the context of those given conditions and assumptions.

"At this point, if possible, controlled experiments are performed that focus on, or isolate, different variables shaping the problem being considered. By comparing the results of these experiments with the character of one's model, one has an opportunity, over time, to correct, eliminate, refine and/or confirm different facets of the model."

"We have before us, Dr. Yardley, three different -- models, I guess -- or theories concerning the faint early sun paradox," noted the defense lawyer. "Is there an experimental way," the lawyer asked, " of deciding which, if any of these models, are an accurate reflection of what happened on Earth during the Archean era?"

"Not really," observed the professor. "Certain experiments might carry various kinds of implications and ramifications for such models that will have to be taken into consideration.

"Experimental results might raise questions about, or pose problems and challenges for, a particular model. Generally speaking, however, what happens is that researchers will merely modify their models in the light of the experimental data.

"Since we, to some extent, are working in the dark concerning what the precise nature of the conditions were during the Archean era, we frequently are limited to saying that different kinds of models are consistent with, rather than proved by, the known facts. Yet, the known facts might be, more or less, equally consistent with quite different models, depending on the assumptions one makes and how one chooses to interpret, and piece together, the available facts in the context of one's model.

"All models are conditional in nature. In other words, the accuracy or reflective capacity of a model, vis-à-vis 'reality', or the facts, or one's field and laboratory experiences, is dependent on the rigor with which, and degree to which, one's assumptions can be shown to be plausible, or justified representations, of the prevailing conditions surrounding some issue or phenomenon.

"Modern scientists cannot recreate the Archean era conditions. At best, we can try to simulate certain facets of what, on the basis of the available data, we believe those Archean era conditions to have been.

"In the light of these simulations, we extrapolate and interpolate backwards to the Archean era. In this fashion, we try to link, as well as we can, our simulations, whether computer or experimental, to the available empirical data and known physical/chemical laws of nature.

"Many models will work in upper and lower boundaries as part of their conditional statements concerning the nature of reality. In other words, if certain variables operate at the upper boundary limits of the model, then, certain things are said to follow. If, on the other hand, these same variables operate at the lower boundary limits of the model, other kinds of things maybe said to follow.

"For example, quite a few simulation experiments in evolutionary theory concerning the Archean era are now, and have been for a number of years, examining the issue of organic synthesis under a variety of prebiotic conditions. The same experiment or simulation will be run a number of times under, say, a variety of conditions involving different atmospheric compositional packages.

"On one experimental run, a particular organic synthesis will be attempted with a methane/ammonia atmosphere. Other runs of the experiment will be done in the presence of, perhaps, different ratios of hydrogen and carbon dioxide gases.

"On the basis of these experimental results, a researcher will reach certain tentative conclusions. For instance, she or he might say: when the composition of the atmosphere consisted of a particular mixture of methane and ammonia, the synthesis went forward at such a rate and with such-and-such an efficiency yield. However, when the same synthesis was attempted with a certain ratio of hydrogen and carbon dioxide, the synthesis either did not occur, or it occurred at a reduced rate and with reduced efficiency yields of such-and-such a nature."

"Dr. Yardley, why don't we, "suggested the defense counsel, "run some data by you and see how you handle it in the context of an evolutionary model? Perhaps, this exercise will help the court and the jurors to get a better feel for some of the issues that are, I believe, at the heart of the present trial."

The professor gestured a willingness to go along with such an exercise. He poured himself a glass of water and waited for the defense lawyer to begin.

"Let's return, for a moment," Mr. Tappin directed, "to the theory which assumes that the world froze, at some point, in response to the faint early sun. You indicated previously that one of the inspirations behind the construction of such a theory was to avoid the potential problems associated with a carbon dioxide dominated atmosphere in the Archean era.

"Presumably, one of these difficulties is that carbon dioxide has the potential to be highly reactive with complex hydrocarbons. As a result, CO_2 will help break the more complex molecules down into less complex and less interesting organic materials as far as the origin-of-life issue is concerned.

"Another difficulty posed by a carbon dioxide dominated atmosphere is the following. Experiments have shown that many kinds of organic synthesis are less likely to proceed or do so in very limited fashion, in such an atmosphere.

"If the Archean era atmosphere were dominated by carbon dioxide -- with very little, or no methane and ammonia -- how would the 'let the world freeze' assumption avoid the ramifications of this kind of atmosphere? In other words, where would the simple hydrocarbons come from out of which more complex hydrocarbons are to be synthesized, and what sources of energy would underwrite this underwater synthesis?"

"If one were to assume," Dr. Yardley responded, "there were little, or no, atmospheric production of hydrocarbons, like hydrogen cyanide (HCN) or formaldehyde CH_2O, then one would have to look to other sources such as carbonaceous chondrites, interplanetary dust particles or interstellar dust clouds, for either these more complex kinds of hydrocarbons or their simpler precursors such as methane and ammonia. Another possibility might be through hydrothermal vents from which hot water, rich in dissolved materials, spills out into the ocean."

"If," posited Mr. Tappin, "the surface of the Earth is frozen over, how do extraterrestrial materials get to the underlying ocean?"

"One possibility," replied the professor "is that these materials might have gone into solution during the earliest stages of ocean formation, prior to the establishing of a frozen zone near the surface of the ocean. Another possibility arises in conjunction with the asteroid impacts that, conceivably, could have provided a mechanism for mixing exogenous/extraterrestrial organic materials lying frozen on the surface with the oceans lying 300 meters below the frozen zone."

"Dr. Yardley, isn't the asteroid impact possibility a bit like dropping a hydrogen bomb on the Antarctic regions and seeing if anything interesting happens?"

"Objection, Your Honor," proclaimed Mr. Mayfield. The question is argumentative."

"Sustained," ruled Judge Arnsberger. "Rephrase the question, Mr. Tappin."

Starting again, the defense counsel asked: "Do we have any good reason to believe the impact of an asteroid sufficiently large to melt 300 meters of ice encircling the globe would have anything but destructive consequences for whatever residual exogenous organic materials might be lying frozen on the surface of the Earth?"

"No, I suppose not," answered Dr. Yardley.

"Is it fair to say, Professor," inquired Mr. Tappin, "that any conjectures concerning what might or might not have survived such a catastrophic event are quite presumptive and arbitrary in character?"

"Yes, I think that would be fair to say," Dr. Yardley acknowledged.

"Would you also agree, Professor," pressed Mr. Tappin, "that in view of the many uncertainties surrounding both the issue of the formation of the Archean era ocean, as well as the uncertain nature of the circumstances and conditions connected to the emergence of the 300-meter frozen zone, any conjectures concerning what had or had not entered into solution prior to the appearance of the frozen zone are equally presumptive and arbitrary?"

"I would have to offer a provisional yes to your question," Dr. Yardley stated."

"What is the nature of your qualifying provision?" inquired the lawyer.

"If," the professor hypothesized, "exogenous or extraterrestrial organic materials were reaching Earth through interplanetary dust particles or by the Earth's passage through interstellar clouds or by means of carbonaceous chondrites, then one would have to consider the possibility that these organic materials might be available to enter into solution should the opportunity arise."

Mr. Tappin briefly left the area of the witness stand and returned to the defense table. He whispered something to his colleague who rifled through

| Origin of Life |

some material on the table and pulled out a sheet of paper that he handed to the defense lawyer.

As the lawyer came back toward the witness, he started to speak. "Professor Yardley," he inquired "are you familiar with a 1993 report by a NASA experimental team concerning the composition of interstellar dust?"

"In general, yes, I am familiar with that report," answered the professor, "but some of its details are rather fuzzy in my mind."

"Let me refresh your memory," offered Mr. Tappin. "The NASA scientists examined a number of star-forming clouds in the Milky Way galaxy."

"In every star-forming cloud, without exception, examined by the NASA team, they discovered that carbon in the form of microscopic diamonds dominated these clouds. In fact, these microscopic diamonds were found in huge numbers and at planetary masses.

"The findings of the research team have been described as a challenge to existing theories of both galactic and star formation. These prevailing theories assumed that interstellar clouds were composed of softer hydrocarbons, somewhat similar to gasoline or candle wax.

"Dr. Yardley, in the light of your previous answer about the availability of different exogenous sources for entering into solution should the opportunity arise, what are the implications of the largely hard-carbon, or microscopic diamond, composition of interstellar clouds?"

"Probably," surmised the professor, "one would have to revise downward one's estimates of the quantities and the kinds of soft hydrocarbons that might have come to Earth by means of its passage through such interstellar clouds. How much these estimates would have to be revised in a downward direction would depend on the extent to which the hard-carbons dominated these interstellar clouds."

Once again, Mr. Tappin returned to his table. On this occasion, his colleague was waiting for him, giving the defense counsel some new material in exchange for the paper in the lawyer's hand.

Approaching Dr. Yardley, the lawyer for the defense stated: "Professor, not too long ago, there was a study that examined the character and composition of a substantial number of extraterrestrial dust grains, which you have referred to as interplanetary dust particles. In more than 50 ice samples taken from a core drilled in the ice of Greenland, and, therefore, representing thousands of years of elapsed time, these researchers found only an extremely tiny amount of amino acids.

"The scientists conducting this experimental analysis concluded that amino acids couldn't have arrived in interplanetary dust particles in amounts that would have any significant bearing on issues concerning the origin-of-life. How would you respond to this finding, Dr. Yardley," asked the defense counsel, "in the light of your previous qualifying provision concerning the availability of exogenous or extraterrestrial organic materials for entering into solution prior to the formation of a 300-meter ice layer caused by a faint early sun?"

"Obviously," the professor noted, "one's estimates again would have to be revised downward. How much, and in what way, would depend on what other kinds of organic materials were found in the analyzed samples."

Shuffling through the papers in his hand, the lawyer selected another document. "Undoubtedly, Dr. Yardley," the lawyer said, "you are aware of the fact that a great deal of the organic materials found in interplanetary dust particles exists in the form of polycyclic aromatic hydrocarbons and amorphous carbon, both of which offer far less promise for the origin-of-life question than do amino acids or purine and pyrimidine nucleic bases. Is my assumption concerning your knowledge correct, Professor?"

"Your assumption is correct," Dr. Yardley replied. "However," he added, "some pathways of synthesis have been proposed that permit one to go from amorphous carbon and polycyclic aromatic carbons to amino acids."

"Yet," countered Mr. Tappin "those pathways are not without their controversial dimensions. Is it not the case Dr. Yardley that other researchers have disputed the proposed pathways of synthesis to which you are referring?" queried the lawyer.

"That's right," the professor admitted.

"Dr. Yardley," inquired Mr. Tappin, "you have testified previously that no one knows, for sure, about the origins of interplanetary dust particles. Is this correct?"

"Essentially, yes," the professor confirmed, "although, as I indicated earlier, some researchers have conjectured these dust particles might have arisen as a result of asteroid-asteroid collisions."

"However, Professor," challenged the defense counsel, "would one be justified in saying there is no proof or evidence in support of such a conjecture?"

"Yes," Dr. Yardley agreed, "one would be justified in saying no hard proof or evidence exists with respect to that conjecture."

"Furthermore," inquired Mr. Tappin, "would one also be justified in pointing out that the conjecture that you have described does not really explain how these dust particles came to contain different kinds of organic materials?"

"Yes," the professor admitted.

"In point of fact, Dr. Yardley," pressed the defense counsel, "given our ignorance about the origins of interplanetary dust particles, we really have no reliable and valid way of projecting backward from current data involving interplanetary dust particles to what might have been going on during the Archean- era?"

"That's correct," replied Dr. Yardley.

"In other words, Professor," continued Mr. Tappin, "we have little or no evidence concerning either the rates of production of interplanetary dust particles or whether the levels of mass influx of such particles that are currently observed would have remained constant across more than 4 billion years, and, therefore, be indicative of the influx of interplanetary dust particles that might have occurred during the Archean era. Is this correct, Dr. Yardley?" inquired Mr. Tappin.

"Yes, I would say so," responded the professor.

"Is it not also true, Dr. Yardley," probed the defense counsel, "that we have no hard, rigorous, reliable data on the amount, or kinds, of extraterrestrial organic material that would have been lost in the Archean era due to: pyrolysis, while in transit through the atmosphere; or, meteorological and geological ablation after air bursting; or, destruction as a result of the effects of shock waves or impact with the Earth; or, ultraviolet decomposition?"

"What you say is true," Dr. Yardley acknowledged.

"Consequently, Professor," concluded the defense lawyer, "the mass influx figures you cited during your direct examination testimony are pure conjecture based on, among other things, the assumption that everything we observe today with respect to interplanetary dust particles has remained essentially unchanged for four billion years. Is this correct?"

"Yes, it is," the professor agreed, "but I would point out that continuity plays a fundamental role in many aspects of the natural laws that govern physical and chemical phenomena."

"Dr. Yardley would you say there are qualitative differences among inorganic chemistry, organic chemistry and biochemistry?" Mr. Tappin asked.

"I would say," answered the professor," there are principles and properties that are shared in common by these disciplines, as well as areas of qualitative difference in which properties and principles that are unique, in a sense, to each of these disciplines do manifest themselves."

"Do you feel one would be justified," inquired the counsel for the defense, "to say the problem of accounting for the emergence of life from prebiotic beginnings is, in part, a reflection of the fact that the transition from organic chemistry to biochemistry involves, at least at the present time, more unresolved problems of a qualitatively different kind than one would encounter in making the conceptual transition from inorganic chemistry to organic chemistry?"

"Yes," confirmed the professor, "at the present time, what you have said is the case. Nonetheless, evolutionary scientists firmly believe the current situation will not last forever.

"We all feel," Dr. Yardley added, "that one of these days a researcher or scientist will demonstrate or discover how the last, unknown steps in the transition from organic chemistry to biochemistry took place. When this happens, the transition from organic chemistry will be no more mysterious than is the transition from inorganic chemistry to organic chemistry."

"Be that as it may, Professor," responded Mr. Tappin, "let me point out the obvious. Evolutionary biologists do not currently have such knowledge.

"More importantly, as far as the present aspect of the cross - examination is concerned, even if evolutionary biologists did possess such knowledge, certain facts still cannot be denied. For instance, let us assume there is some continuous set of chemical principles that allows one to make the transition from organic chemistry to biochemistry through purely natural processes.

"Nevertheless, there still are phenomena that occur in biochemical systems that do not take place in the reactions of systems that are organic but non-biochemical in character. Is this not so, Professor?" inquired Mr. Tappin.

"Yes," confirmed the biologist.

"In brief, Dr. Yardley," the lawyer summarized, "things do not always remain the same over time. If they did, we wouldn't be having this debate about why post-prebiotic times exhibited properties that were not present in prebiotic times.

"What happens now is not necessarily what was happening in the past. Moreover, what happened in the past is not necessarily what is happening now.

"This general principle, if you will, is demonstrated by the qualitative differences between biochemical processes compared to purely organic ones. This principle also might be demonstrated by possible differences in the rates of mass influx of interplanetary dust particles between today and 4 billion years ago.

"Would you agree, therefore, Dr. Yardley" asked Mr. Tappin, "that although we would expect the same conditions to exhibit the same properties over time, we cannot expect different circumstances automatically to lead to the same manifested properties? In fact, isn't the problem with which we are confronted in this matter of the mass influx rates of interplanetary dust particles, a variation on this theme?

"We need to determine the precise nature of the conditions under which an individual is justified in concluding that the things that are observed today are the same as what would have been occurring in the Archean era. Is this not part of the problem before us, Professor?"

"Yes, I think I could live with your characterization of things," Dr. Yardley stipulated.

Mr. Tappin began to speak and was interrupted by Judge Arnsberger. "Mr. Tappin, I'm sorry, but in view of the lateness of the hour, I feel we would be well advised to adjourn these proceedings for the day.

"I hope you will agree that the present time seems to offer a natural point of transition in your cross-examination. In any case, you will be able to pick things up again at 10:00 a.m. tomorrow morning."

Turning her attention to the jury, she said: "Please remember, ladies and gentlemen, my previous instructions to you. You are prohibited from discussing this case either with fellow jurors or with others whom you might come into contact.

"Court is adjourned until 10:00 a.m., Thursday morning," announced the judge. Her gavel fell in confirmation of her words.

An Ocean of Difficulty

Going through the papers in his hand, the defense counsel removed several sheets. Walking over to his table, he returned the unwanted sheets to his colleague.

Standing in front of the defense table, Mr. Tappin said: "Professor, in your direct examination testimony, you indicated, I believe, that the Murchison meteorite contained 6 amino acids similar, in most respects, to amino acids occurring in living organisms. In addition, 12 other kinds of amino acids not found, as far as is known, in living organisms on Earth also were discovered in the Murchison meteorite. Is my recollection of this testimony correct?" asked the lawyer.

"Yes," Dr. Yardley confirmed.

"To the best of your knowledge, Professor," Mr. Tappin inquired, "has any recovered meteorite ever contained all twenty of the amino acids found in living organisms on Earth?"

"Not to my knowledge," the professor answered.

"Furthermore," continued the defense counsel, "you testified that the amino acids found in the 200,000-year old meteorite in Antarctica had optical properties that were opposite to the ones displayed by amino acids found in Earth organisms. Is this correct, Dr. Yardley?"

"Yes, it is," the professor responded.

"In addition, Dr. Yardley, I believe you stated earlier that in most cases outside of biological systems, amino acids tend to form racemic mixtures in which there are roughly equal numbers of left- and right-handed optical isomers. Is my understanding correct in this respect?" Mr. Tappin inquired.

"Yes," said the professor.

"Moreover, previously, you testified that only 5-6 percent of meteorites consist of carbonaceous materials, and organic materials constitute only a small part of this carbonaceous subset of meteorites. Is this right?"

"Correct," affirmed the professor.

"Finally, Dr. Yardley, isn't it the case that most of the organic material found in meteorites such as Murchison exists in the form of a complex kerogen-like polymer that is poorly defined and consists of a variety of aromatic groups, monocarboxylic acids and aliphatic hydrocarbons? In fact, isn't it true, Professor, that only a very small fraction ... measured in parts per million ... of the organic material found in meteorites contains molecules, such as purines and amino acids, which have any potential relevancy to issues concerning the origin-of-life?"

"That is right," the professor indicated.

"Well, Dr. Yardley," the attorney stated "if we factor in all of the foregoing possibilities, we seem to be left with very uncertain, and possibly negligible, amounts of usable organic compounds from exogenous sources. In other words, given that organic materials form only a tiny portion of an already small subset of meteorites, and given that many of these exogenous organic materials exist in forms, or as kinds, which are not used by Earth organisms, and given that a considerable amount of this organic material might be destroyed through pyrolysis, hydrolysis, photolysis or impact, and given that we really don't know the rate or mass of carbonaceous chondrite influx during the Archean era, are not any statements about the amount and kinds of useable exogenous organic materials that arrive, and survive, very speculative and arbitrary?"

"Yes, I suppose so," Dr. Yardley admitted.

"Earlier," Mr. Tappin noted, "you mentioned, briefly, the possibility that hydrothermal vents might have played a role in the 'let the Earth freeze' model that arose in response to, among other things, the faint early sun paradox. Would you expand on this a little?" the lawyer requested.

"Some people," the professor said, "began to look seriously at hydrothermal vents as a possible locus for the origin-of-life when, a few years ago, rather extensive ecosystems were discovered to have developed around some of these vents. These ecosystems consisted of many exotic sorts of organism, including blind shrimp and giant tube worms.

"The food chains of these ecosystems were rooted in various kinds of microorganisms. These microorganisms were sulfur-eating life forms.

"Thermophilic, or heat-loving, microbial organisms also have been found living in the steam bath-like conditions of the hot springs at Yellow Stone National Park. In general, however, no one has discovered life forms on Earth capable of surviving in temperatures above 112 degrees Celsius."

"My understanding, Professor," indicated the lawyer "is that these organisms are capable of living under such conditions because they possess specialized proteins that allow them, among other things, to dissipate heat. Apparently, there also are proteins in various species of cold water fish capable of binding to, and controlling the growth of, ice within the organism, and, as a result, helping the organism adapt to cold water conditions. Is this correct?"

"Yes," replied the professor. When he saw the defense lawyer signaling him to continue on with his discussion, he said: "Some researchers hypothesized that life might have originated with thermophilic organisms.

"Other scientists have hypothesized that life originated elsewhere. In time, however, these organisms might have migrated to the hydrothermal vents in order to seek resources exuded by the vents or as a protection from the extraterrestrial bombardment of the Archean era Earth, or, maybe, both.

"Presumably," reflected the defense counsel, "if organisms migrated to the vents, then, regardless of whatever forces drove organisms to, or induced them to seek out, these hydrothermal vents, nonetheless, in order to survive these organisms would have to be adapted, in some minimally feasible fashion, to the thermal conditions of the vents. Is this not so, Dr. Yardley?"

"That's right," acknowledged the professor.

"But, the process of migration presupposes the existence of such organisms and assumes the existence of such adaptive capabilities. So, we are getting ahead of ourselves.

"Has anyone," Mr. Tappin asked, "devised a plausible theory of how life would have originated in the vicinity of the hydrothermal vents?"

"Not really," replied the professor.

"Dr. Yardley, in your direct examination testimony concerning the period of core differentiation of the Earth, you indicated some scientists believed the Earth's crust would have been relatively fragile at that time, and, therefore, conducive to the formation of these hydrothermal vents. Is this right?"

"Yes," responded the professor.

"Does the water in the ocean remain relatively static, or does it circulate?" Mr. Tappin asked.

"The water in the oceans of our day circulates extensively," the professor reported. "In fact, we believe any given volume of water eventually will circulate through every portion of the ocean."

"What about the Archean era ocean?" inquired the lawyer?

"I think the same scenario probably was the case," offered Dr. Yardley. "Between tidal forces and convection currents, of one sort or another, a circulatory system of some kind likely would have been present."

"If my information is correct," Mr. Tappin stated, "the temperatures associated with hydrothermal vents are in the vicinity of 350 degrees Celsius. What would be the effect," queried Mr. Tappin, "of hydrothermal vents on complex hydrocarbons that had dissolved in ocean waters and were brought into contact with these vents through the process of circulation?"

"A lot would depend on the extent, length and character of the contact," replied Dr. Yardley. "In general, the more direct, the longer, and the more extensive such contact, the more likely would be the tendency of any given complex hydrocarbon to denature or decompose."

"Would one be justified in arguing," asked the defense counsel, "that given some unknown number of hydrothermal vents on the bottom of the Archean era ocean, then the formation of a 300-meter ice layer above the ocean, due to the effects of a faint early sun, would not necessarily offer long-term stability to complex hydrocarbons that had, in one way or another, arisen?"

"As long as the molecules were able to stay in cold or cooler waters," Dr. Yardley pointed out, "then, their average lifetimes probably would be enhanced to some degree. On the other hand, to whatever extent such molecules could not stay in cold or cooler conditions, then the average length of life for such molecules would be decreased as a function of the different kinds of forces of decomposition, including temperature, to which these molecules were subjected.

"For example, one scientist has studied the effects of heat energy on the amino acid alanine. This molecule is one of the more stable amino acids.

"The researcher found that at a temperature of twenty-five degrees Celsius, the mean life of alanine is estimated to be 10^{11} years. Yet, the mean life of this molecule is calculated to be just thirty years in length when the temperature is raised to 150 degrees Celsius.

"Less stable amino acids will break down more readily at such temperatures, and, therefore, they will have even shorter mean life times than alanine. In fact, less stable amino acids might begin to break down at temperatures somewhat lower than 150 degrees Celsius.

"Generally speaking, the more complex a hydrocarbon, the more unstable it tends to be in the presence of heat. For instance, proteins, DNA, and RNA all tend to denature and decompose when exposed to sufficient amounts of heat much more readily than might be the case with their component parts."

"Dr. Yardley," said the defense counsel, "I presume the aforementioned effect of heat on complex organic molecules would remain the same whether one is talking about hydrothermal vents or elevated surface temperatures caused by a super greenhouse effect. Is this presumption correct?"

"Yes, of course," remarked the professor.

"Therefore," Mr. Tappin observed, "all three theories that have been proposed as possible ways of resolving the faint early sun paradox, face, each in its own way, a potential problem with respect to decomposition of complex hydrocarbons as a result of potentially prolonged exposure to heat energy, either in relation to hydrothermal vents or to enhanced greenhouse effects. Would you agree with this assessment of the situation, Dr. Yardley?"

"In broad terms, I suppose so," answered the professor."

Looking briefly at the papers in his hand, Mr. Tappin walked toward the witness stand. When he was a few feet away, he came to a standstill.

"Professor, earlier you testified that scientists believe there was little or no free oxygen in the early Archean era atmosphere. Given," postulated the defense counsel, "all the talk these days about holes in the ozone layer and how ozone absorbs ultraviolet radiation, and, in the process, protecting living organisms from the destructive effects of such radiation, I was wondering what the situation would be in the Archean era. More specifically, would the faint early sun lessen the presence of ultraviolet radiation?"

"Oddly enough," Dr. Yardley began, "although the overall, net luminosity of the early sun was lower than today's sun, nonetheless, on the basis of astronomical observations of young stars comparable to our early sun, the ultraviolet radiation of the early sun is considered to have been greater than is the case with our present sun. Consequently, in the absence of oxygen, the ultraviolet effect would be more pronounced than it is today, even in those areas, such as the Antarctic, where the ozone hole has grown to such a disturbing size."

"Where does ozone come from?" Mr. Tappin asked.

"When free oxygen is available," Dr. Yardley explained, "ultraviolet radiation tends to split oxygen molecules into separate atoms of oxygen that are quite unstable. These unstable atoms of oxygen will combine with oxygen molecules to produce O_3 or ozone.

"Studies have indicated there was no appreciable presence of atmospheric oxygen until sometime between 2.1 and 2.03 billion years ago. As a result, between 4.55 billion years ago, and 2.1 billion years ago, there would have been no way for ozone to be manufactured in the Archean era atmosphere."

"What are the ramifications," Mr. Tappin inquired, "of this combination of enhanced ultraviolet luminosity, as a result of the faint early sun, and the absence of ozone, due to the absence of oxygen, as far as the development of increasingly complex hydrocarbons is concerned?"

"Ultraviolet light," replied the professor "is like most forms of energy. They are all two-edged swords.

"In the right amounts and for the right length of time, energy is capable of bringing about many kinds of chemical reactions among organic molecules. In the wrong amounts and for the wrong length of time, energy can be quite destructive in its effects upon hydrocarbon compounds.

"In limited doses, ultraviolet radiation can help underwrite, among other things, the synthesis of a wide variety of organic molecules. Beyond a certain limit, however, such radiation begins to have an adverse effect, even on those compounds that, originally, it might have had a hand in helping to synthesize.

"Photolysis refers to the breakdown or decomposition of materials by the action of light. Prolonged exposure to ultraviolet radiation brings about photolysis.

"These remarks notwithstanding, the results of photolysis sometimes can bring about reactions that have a potential, under the right circumstances, for building more complex hydrocarbons. In other words, the products of photolysis might recombine with other organic materials.

"For example, one team of researchers observed that when methane gas is subjected to photolysis, methyl (CH_3) and methylene (CH_2) radicals are produced. Subsequently, these two radicals were observed to enter into reactions that resulted in heavier hydrocarbons.

"These researchers calculated that the equivalent of one bar or atmosphere of methane gas could have been polymerized by means of ultraviolet radiation over a period of some 10^6 to 10^7 years ... in other words, between one and ten million years. They further proposed that such heavier hydrocarbons would have precipitated out of the atmosphere and formed a layer of hydrocarbons on the surface of the Earth measuring anywhere from one to ten meters in thickness."

"Dr. Yardley," interjected the defense council, "in the light of our previous discussion about the nature of the atmospheric composition of the Archean era, couldn't one respond to the findings of this methane photolysis research in several ways? For example, if the Archean atmosphere were methane-dominated, this finding might have some value in origin-of-life scenarios, but if the Archean -atmosphere consisted of little or no methane, their finding is meaningless as far as the origin-of-life issue is concerned. Would you agree with this assessment of the situation, Dr. Yardley?"

"Not entirely," the professor indicated. "Even if there were little methane in the atmosphere, the synthesis of important precursors ... such as hydrogen cyanide, formaldehyde, and, maybe, a few amino acids, still is possible.

"A great deal would depend on the ratio of hydrogen (H_2) to carbon dioxide gas (CO_2) that existed in the Archean -atmosphere.

"As I testified previously, if the ratio were about 2, then, some researchers feel this kind of atmosphere would have reducing properties comparable to a methane-dominated atmosphere.

"As the ratio of hydrogen gas to carbon dioxide drops, the production efficiency by ultraviolet light also will drop. As one approaches a ratio of, say, one-tenth of hydrogen to carbon dioxide, then production efficiency by ultraviolet light is calculated to drop by at least two magnitudes or by a factor of around 100.

"Researchers suggest hydrogen might have arisen throughout ... gassing from Archean era volcanoes. Hydrogen also might have been generated through the photo-stimulated reduction of ferrous iron in the photic zone of the ocean."

"Doesn't this photo stimulated reduction of ferrous iron assume," observed the defense counsel," that the surface of the Earth has not been frozen over due to the effect of the faint early sun?"

"Obviously," the professor responded.

"In addition," continued the lawyer, "doesn't the temperature of the exosphere, some 400 miles above the Earth, have to be factored into the equation concerning hydrogen? Doesn't the rate at which hydrogen escapes from the Earth's atmosphere increase as the temperature of the exosphere rises?"

"Yes, that is correct," acknowledged the professor.

Turning over one of the papers in his hand, the defense counsel ran the fingers of his right hand down the page. At a point near the bottom of the page, he stopped and inquired: "Are you familiar with Shimizu's study on exospheric temperatures in a methane dominated Archean era atmosphere?"

"Vaguely, yes," Dr. Yardley answered.

"Shimizu had concluded," reported the lawyer, "that a methane dominated Archean era atmosphere would have had an exosphere whose temperature exceeded 1300 degrees Kelvin or more than 1000 degrees Celsius. The study suggested these temperatures would have made an atmosphere of such composition very short-lived.

"If one were to assume," Mr. Tappin postulated, "that a super greenhouse effect in a carbon dioxide- dominated atmosphere also were capable of generating comparable kinds of exospheric temperatures, then might one conclude, with some degree of justification, that there could be a relatively high rate of exodus of hydrogen from such an atmosphere?"

"Possibly," Dr. Yardley offered.

"Moreover," Mr. Tappin countered, "irrespective of the kind of atmosphere in which organic materials might have arisen by means of ultraviolet synthesis, if such organic materials were to continue to remain in the same exposed condition to ultraviolet radiation, then they will, after a time, begin to break down or decompose through the process of photolysis. Is this right?" inquired Mr. Tappin.

"That's pretty much the upside and the down side of things," answered the professor.

"Let us assume," proposed the defense counsel that a methane-dominated atmosphere, or its hydrogen/carbon dioxide equivalent, existed. Let us further assume that the equivalent of one atmosphere of methane gas, or its hydrogen/carbon dioxide equivalent, was polymerized to more complex hydrocarbons through ultraviolet photolysis over a period of some 1 to 10 million years.

"Despite allowing such assumptions as given, one still would have to consider the following possibility. The one to ten meters of organic material that we are assuming had precipitated out would now be subject to one to ten million years of further photolysis, not to mention possible hydrolysis, and, depending on surface temperature, pyrolysis. Is this about right, Professor?"

"More or less," Dr. Yardley said.

"In addition," continued the defense counsel, "if there were an extraterrestrial event of sufficient magnitude to vaporize the ocean, or vaporize the photic zone, or the size of the Yucatan meteorite, then, the one to ten-meter layer of hydrocarbon material that has been postulated by some, would be, shall we say, history. Would you agree with this?"

"Given your premise, that conclusion follows," admitted Dr. Yardley.

Referring briefly to the paper in his hand, Mr. Tappin asked: "Professor, would one be correct in stating that only a small fraction of the light energy coming from the sun is in the form of ultraviolet wavelengths that are sufficiently small to be capable of being absorbed by molecules such as H_2O, CO_2, CH_4, and NH_3?"

"Yes," agreed the professor.

"Would one also be correct," inquired the lawyer, "if one said the following: when more complex molecules are formed, then, the absorption profile or spectrum of these molecules shifts in the direction of longer wavelengths where a great deal more energy is available from the light being radiated from the sun?"

"Again, yes," the professor affirmed.

"Dr. Yardley," continued the defense counsel, "do most of the relatively low wavelength ultraviolet photochemical reactions take place in the upper or lower atmosphere?"

"The upper atmosphere," responded the professor.

"Is it possible," queried the lawyer, "that the compounds which formed in the upper atmosphere through low wavelength ultraviolet photochemical reactions are now vulnerable to photolytic decomposition in relation to a broader range of energies as the absorption spectrum of these more complex compounds moves in the direction of longer wavelengths?"

"Yes, this is a possibility," the professor acknowledged.

"In other words, Dr. Yardley," the defense counsel summarized, "a variety of compounds could have been synthesized in the upper atmosphere by means of low-wavelength ultraviolet photochemical reactions, and, then, these newly formed compounds could have been decomposed through the photolysis brought about by longer wavelength ultraviolet radiation to which these compounds had become susceptible by virtue of their greater complexity, and, this all could take place before the organic materials ever reached the ocean or surface of the Earth. Isn't this a very real possibility, Professor?"

"Yes, it is," Dr. Yardley stipulated.

"Seemingly," Mr. Tappin suggested, "there is something of a race between two opposing forces here: photolytic production of compounds and photolytic decomposition of organic materials. Which of these two forces dominates in a given context will significantly shape what does and does not get to the ocean. Is this correct Dr. Yardley?"

"I would say so," the professor confirmed.

Once again, Mr. Tappin went to the table for the defense and exchanged the papers in his possession for ones being offered by his colleague. Turning back toward the witness, the lawyer said: "Dr. Yardley, in your direct examination testimony concerning the coupling of shock wave energy to hydrocarbon synthesis, you cited a number of figures."

Reading aloud from the papers in his hand, the lawyer summarized the material. "One, meteorites entering the atmosphere with a mass between $10^{-14} - 10^2$ grams would generate, collectively, about 1.8×10^{15} joules per year. Two, carbonaceous chondrite airbursts of objects that had a radius less than, or equal to, 300 meters would generate, collectively, approximately 1.5×10^{14} joules per year. Three, the post-impact vapor plumes of meteorites striking the Earth's surface would produce – collectively -- about 6×10^{17} joules per year. Are these figures correct, Dr. Yardley?" asked the defense lawyer.

"Yes," the professor indicated.

"What sort of a conversion factor is used to come up with these figures?" Mr. Tappin inquired. "In other words, what percentage of the total impact energy actually is believed to be directed toward, or available for, shock synthesis?"

"The conversion factor," replied the professor, "would be a function of the kind of assumptions one made in developing the thermochemical model one used to calculate energies, efficiencies and so on. The amount of total energy that is capable, potentially, of being converted to synthesis reactions starts at about twenty to thirty percent and works its way downward from there depending on the factors being taken into consideration."

"Presumably then, Dr. Yardley," remarked the lawyer "the figures you have cited are not cast in stone. The actual energies that might be directed toward synthesis reactions might be less, perhaps even, considerably so, than the figures you have cited. Would you agree with this?"

"To a certain extent," the professor responded. "At the same time, these figures are not randomly pulled out of a hat. They are the end result of quite a bit of rigorous reflection and take into consideration a great deal of scientific knowledge."

"I'm sure," admitted the lawyer, "that what you say is true, Dr. Yardley. However, the same thing could be said with considerable justification at almost every stage of science for the past several hundred years, and, yet, despite this, models have changed and calculations have been revised. Isn't this so, Professor?"

"I suppose so," replied the professor.

"Dr. Yardley, if one varied the value of atmospheric pressure in one's model, how would this affect calculated energy values with respect to meteorite influx?" Mr. Tappin wondered.

"Within certain limits," the professor suggested, "increasing the atmospheric pressure would help to aerobrake incoming objects. This would decrease, to some extent, the velocity of these objects and, consequently, would tend to affect the total amount of impact energy manifested during the passage of the meteorite through different parts of the atmosphere."

"Would it be fair," Mr. Tappin inquired, "to say that we do not know, in any of the three sets of figures, how the energy is distributed over time or across space? In other words, Professor, wouldn't some days, hours or minutes receive disproportionate amounts of these yearly allotments of energy relative to other days, hours and minutes? Similarly, wouldn't it be the case that the billions of cubic miles of atmosphere that surround our planet will not all receive an equal and even distribution of the yearly allotments of energy for any of the three ways of generating energy?"

"This is likely to be the case" Dr. Yardley answered.

"Is it not also true," asked Mr. Tappin, "that the efficiency of shock wave synthesis decreases in relation to increases of impact energy? In other words, isn't it true that the yield of organic materials per unit of impact energy decreases as the impact energy increases?"

"That's correct," responded the professor.

"Consequently," the lawyer reasoned, "there will be variable yields of organic materials due to shock synthesis as a function of the impact energy for any given set of spatial and temporal coordinates. Would you agree with this, professor?"

"I would," Dr. Yardley acknowledged.

"In the light of the foregoing considerations," stipulated the defense counsel, "could one argue in the following fashion? Is it conceivable that some of the shock wave energy created in the atmosphere by microscopic-sized meteorites might generate shock wave energy in such a way that either: (a) the pattern of energy distribution across space might not be sufficiently concentrated in any one area to bring about organic synthesis; or, (b) that the precise character of the pathway of the shock wave created by the passage of the microscopic -sized meteorite might not engage any molecules capable of being synthesized with the available energy?"

"Yes, I guess such a scenario is conceivable," acknowledged the professor, "but your description of the situation is very vague?"

"Yes, it is, Dr. Yardley. It is vague in precisely the same way as when one says that meteorites ranging in size from 10^{-14} to 10^2 grams enter the atmosphere and generate 1.8×10^{15} joules of energy per year, and no precise indications are given as to what is happening at any given moment in time and space.

"To speak in terms of yearly energy yields can be quite misleading. We have no way of knowing whether, at any given point in time and space, we have too much energy, which will tend to decrease shock processing yields, or too little energy and, therefore, not enough to generate sufficient energy of activation for a particular synthesis to occur. Would you agree with this, Dr. Yardley?"

"Yes," replied the professor. "Citing yearly energy production in isolation can be misleading. Knowing the particulars of this energy distribution across time and space would be much more important and helpful."

Glancing at the material in his hand, Mr. Tappin inquired: "Speaking of particulars, Dr. Yardley, isn't it true that the shock waves from the post-impact vapor plumes you mentioned don't match up well with the atmosphere as far as how their energy is distributed across space and time? In other words, don't these post-impact plumes rise considerably above the atmosphere and, as a result, release a great deal of their energy outside of the portions of the atmosphere where chemical synthesis is likely to take place?"

"That's right," the professor acknowledged. "I believe some researchers have suggested that, perhaps, only as little as one-sixtieth of the energy from these vapor plumes might be distributed in the atmosphere and, therefore, be available, potentially, for synthesis reactions."

"Therefore, would one be correct in assuming," Mr. Tappin asked, "that not all of the energy generated by shock waves will necessarily be coupled with certain molecules in the atmosphere to produce various synthesized hydrocarbons, and, therefore, some of the available energy will be lost?"

"Yes," replied the professor. "Generally speaking, however, researchers speak about the energy yield in such contexts. In other words, rather than talking about the amount of energy that might or might not be lost, researchers average the mass of the synthesized material across the available energy and, consequently, speak in terms of the amount of material yielded per unit of energy.

"Nevertheless, each unit of energy does not necessarily participate in the synthesis of some given amount of organic material. Energy yield constitutes a relational index of sorts that links the totality of materials synthesized with the totality of energy available for such synthesis.

"For instance, previously I had discussed certain laboratory experiments that investigated the amount of energy generated by rapidly-expanding gases in shock-wave tubes. Researchers involved with these studies found that 332 nanomoles of hydrogen cyanide (HCN) was the yield, on average, for each joule of shock wave energy present in the tube.

"A nanomole is one billionth of a mole, and a mole is the amount of a given substance that is equivalent to the molecular weight of that substance as expressed in grams. So, 332 nanomoles per joule constitutes, in and of itself, a very small quantity of synthesized material, but when multiplied by the total energy of the shock-wave, the overall quantity becomes much larger."

"Yet, isn't it true," noted Mr. Tappin, "that energy yield figures will vary from one atmospheric composition to the next? For instance, given the same magnitude of shock-wave energy in a reducing and a neutral atmosphere, the energy yield of, say, HCN tends to be significantly higher in a reducing environment as opposed to a non - reducing atmospheric environment. Is this correct?"

"Yes," the professor agreed. "In general, the mass of organic materials capable of being shock-synthesized in a given kind of atmosphere - that is, the organic synthesis efficiency - is very dependent on the compositional character of the atmosphere being considered.

"For example, in one thermochemical model of shock-synthesis to which I alluded to earlier," the professor pointed out, "a methane dominated reducing atmosphere is calculated to give a production efficiency of about $10^{17.5}$ molecules of HCN, hydrogen cyanide, per joule of energy. In addition, this was accompanied by the production of a few other kinds of simple hydrocarbons such as C_2H_2 and C_2H_4.

"However, in a carbon dioxide/nitrogen-dominated atmosphere, the production efficiencies for HCN were calculated by this thermochemical model to be approximately $10^{7.5}$ smaller than in the reducing atmosphere. On the other hand, the production efficiencies for formaldehyde (H_2CO) in the neutral atmosphere were calculated to be roughly comparable to what would be obtained in a reducing atmosphere."

"Would you agree, Dr. Yardley," the defense counsel asked, "that thermodynamic calculations might tell one whether or not certain kinds of reactions are possible, but they can say nothing about whether such reactions will occur, nor anything about at what rate they will proceed, nor the path that will be taken by such a reaction?"

"That's right," the professor confirmed.

Mr. Tappin turned the papers in his hand over and began examining the other side. "Dr. Yardley, doesn't the energy yield index you were talking about previously vary with the nature of the energy source involved in any given case?"

"Yes," replied Dr. Yardley. "In experiments with artificial lightning, for example, researchers observed an energy yield of about 3 nanomoles of HCN per joule of energy in a methane-dominated atmosphere versus an energy yield of, approximately, 1000 times less than this in an atmosphere dominated by carbon dioxide."

"Professor, is the magnitude of the energy associated with artificial lightning the same as is associated with natural lightning?" asked the lawyer.

"No, there is a considerable difference between the two," the professor indicated. "In general, natural lightning is far more powerful than artificial lightning."

"Consequently, given what you have said previously," posited the defense counsel, "one would expect the shock-processing yield per unit of energy for natural lightning to be less than that of artificial lightning since the yield per unit of energy decreases as the energy of the impact increases. Is this correct, Professor?"

"This would be consistent with what has been said," Dr. Yardley confirmed.

"Would one be shaky or firm grounds" Mr. Tappin inquired, "if one were to argue that just as there might be various kinds of organic synthesis that occur in the shock-wave wake of meteorites and lightning, so too, meteorites and lightning also can cause the decomposition of materials through pyrolysis and so on?"

"Fairly firm grounds, I would imagine," answered the professor."

"Therefore," stated Mr. Tappin, "in somewhat analogous fashion with respect to ultraviolet radiation, in those circumstances when shock-wave energy is present, there are forces, both of synthesis as well as decomposition, which are taking place, so to speak, side by side. Are we not dealing here, Professor, with the fact that what is being given with the hand of synthesis, is, to some extent, being taken away by the hand of decomposition?"

"Yes," replied Dr. Yardley. "This seems to be the case."

"Apparently," observed the defense lawyer, "we require some kind of 'net energy yield' figure. We need to be able to determine whether the upside, or the down side, of photolysis, pyrolysis, hydrolysis and other factors is dominating any given feature of the Archean era world. Would you agree with this?"

"Yes, I do," the professor affirmed, "but this is easier said than done."

"Dr. Yardley, in the initial origin-of-life experiment performed by Stanley Miller, methane, ammonia, hydrogen and water vapor were used to simulate what was believed, at least at that time, to be the composition of the Archean era atmosphere. In addition, a continuous spark discharge was applied to the gaseous mixture in order to simulate the presence of lightning in a prebiotic world.

"After letting this experiment run for a number of days, the materials synthesized during the course of investigation were examined. Is this very general description of Miller's experiment accurate for the most part?"

"Yes," the professor indicated.

"We know," Mr. Tappin continued, "that questions have been raised by other scientists and researchers in relation to whether or not the Archean atmosphere actually was predominately methane/ammonia in character. I was wondering, however, about the spark discharge aspect of the experiment.

"What was the magnitude of the electrical discharge?" asked the defense counsel.

"Somewhere around two to four watts, I believe," the professor offered.

"Correct me if I am wrong, Dr. Yardley," requested the lawyer, "but I'm not familiar with any 2-4-watt lightning discharges that run continuously for several days. Are you?"

"No," smiled the professor.

| Origin of Life |

"Dr. Yardley," stated the defense counsel, "one might assume that continuous spark discharges from a coil are different in character from lightning bolts and their associated shock waves. Would such an assumption be correct?"

"Well," the professor replied, "the two certainly involve different magnitudes of energy, but the underlying physics is essentially the same. Of course, lightning would not be continuous, but the sparking mechanisms used in the experiments are continuous in nature."

"Would one be unreasonable," Mr. Tappin queried, "to expect different sorts of outcome if one, first, were to expose a certain mixture of gases to a single bolt of lightning and, then exposed the same kind of gaseous mixture to a continuous spark of 2-4 watts for a number of days?"

"No," replied the professor, "probably not, but neither would one be unreasonable if one were to anticipate some degree of overlap in the product outcomes of the two experiments. For instance, both the 2-4-watt spark discharge as well as the lightning bolt might generate some amount of hydrogen cyanide (HCN) in the right kind of atmosphere."

"Can one assume," Mr. Tappin inquired, "that if lightning occurs in the Archean era atmosphere, one will observe amino acids being formed as occurred in the Miller experiment?"

"No, one couldn't assume this," the professor remarked. "In point of fact, the Miller experiment involved a continuous circulation of the gases through the chamber where the electrical spark was being discharged.

"Initially, molecules like formaldehyde and hydrogen cyanide would be synthesized. Then, as these molecules along with the original gases continued to be exposed to the electrical discharge of the spark chamber, slightly more complex molecules in the form of amino nitriles would have been formed.

"Amino nitriles plus water plus continued exposure to the electrical discharge yielded amino acids such as alanine or glycine plus ammonia. There also were a variety of amino acids synthesized that do not occur in any of the biological organisms with which we are familiar."

"Professor Yardley, you have previously testified," Mr. Tappin indicated, "that extremely tiny amounts of hydrogen cyanide were formed when artificial lightning was discharged in a methane-dominated gas mixture, and, you also have testified that hydrogen cyanide was generated during an early stage of Miller's original spark-discharge experiment. Is this correct?"

"Yes, it is," Dr. Yardley remarked.

"You also testified that formaldehyde (H_2CO) is generated during one of the early stages of the Miller experiment. Were there any findings concerning the production of formaldehyde in the artificial lightning studies of which you are aware?"

"In the limited studies that have been carried out," replied the professor, "no formaldehyde formation has been detected. Furthermore, as far as I know, even the figures that come from purely theoretical thermochemical calculations indicate no formaldehyde formation is to be expected in relation to lightning discharges, whether these are artificial or natural."

"Yet, Dr. Yardley, in the Miller experiment, the formaldehyde produced by spark discharge combined with the hydrogen cyanide produced by spark discharge and entered into reaction with ammonia, one of the gases in the supposedly simulated Archean -atmosphere of the experiment, and all of this resulted in the formation of amino nitriles. Is this correct, Professor?"

"That's right," Dr. Yardley agreed.

"Isn't it also the case, Professor," the lawyer inquired, "that researchers believe ion-molecular and free radical reactions, rather than lightning-like shock synthesis, are the essential processes involved in synthesis reactions in spark discharge experiments?"

"Yes," acknowledged the professor.

"In what sense, then, Professor," asked the defense counsel, "can one say the Miller experiment is a simulation experiment, given that it probably simulates neither the atmospheric composition of the Archean era nor the character of lightning discharges, nor the products of lightning discharges, and given that, previously, you have suggested amino acids were formed in the ocean through a Strecker-like synthesis process rather than in the atmosphere through electrical discharges?"

"As far as the features that you have pointed out," replied the professor, "the Miller experiment really isn't much of a simulation experiment. What it does show is this: if one continuously exposes a gaseous mixture of the right molecular composition to an electrical discharge of a certain magnitude, one can generate a series of chemical reactions that will culminate in the formation of complex hydrocarbons that have implications for origin-of-life issues.

"One would have had, perhaps, a closer simulation of certain aspects of actual Archean era prebiotic conditions if one had removed the products of each activation step so that the products of one set of reactions would not have been exposed to the energy source a second time. This process of removing synthesized reaction products at each step of the experiment would have simulated, to a degree, the passage of molecules, synthesized in the Archean era atmosphere, to the ocean, where they would have been protected from further exposure to various forms of energy impinging on the atmosphere."

"Dr. Yardley, wouldn't one have an even better kind of simulation," Mr. Tappin asked, "if one exposed the products of each reaction step to all of the conditions and forces that could have acted upon them in an Archean era context, including the ones that could decompose or destroy such products?"

"Yes, I guess so," agreed the professor, "but there is a practical limit to what can be accomplished in the laboratory."

"Yet," the lawyer countered, "wouldn't you agree that the more we will allow such limitations to distance us from the actual conditions of the world, then the more we will introduce distortions, biases and error into our experimental procedure? Moreover, wouldn't these kinds of distortions skew our capacity to interpret accurately the significance of what our experiments have to say about the nature of the physical world, whether in relation to the natural phenomena of our present day, or those of the Archean world?"

"I would agree," responded the professor, "that we must continuously seek to probe the limitations of our current experimental methods in order to devise, where possible, better experiments and procedures that will permit us either to overcome, or compensate for, such limitations."

"Professor, one could agree with every word you have just said," Mr. Tappin maintained, "but your words do not address or answer the problem before us. To what extent, do the simulations, calculations, estimates, experiments, conjectures, hypotheses and models of prebiotic, evolutionary theory reflect the conditions, forces, processes and dynamics of the Archean era Earth?

"On the basis of testimony that you have given, Dr. Yardley, Miller's experiment doesn't simulate, or emulate, the Archean era world in any way. What his experiment establishes is this: if you do certain things, certain things happen.

"Given that the things that the experiment has done are not necessarily what happened in the Archean era world, then, the fact certain things have been observed to happen might be interesting, intriguing or suggestive, but they don't necessarily shed any light on what actually took place during prebiotic times. Isn't this so, Dr. Yardley?"

"I would agree," the professor admitted, "that the Miller experiment, or others like it, don't prove what happened in the Archean era world. Nonetheless, such experiments generate data that can be incorporated into a process of theory construction that permits the scientific community, over time, to understand, in a consistent, rigorous fashion, a wider and wider body of technical information about an array of interconnected physical and chemical phenomena."

"Yes, Dr. Yardley," Mr. Tappin said, "but the question is this: to what extent does this condition of understanding a wider and wider body of technical information about an array of interconnected physical and chemical phenomena in a consistent, rigorous fashion provide one with a correct understanding of what actually did happen during the Archean era... rather than with just an understanding of what might have happened or what could have happened if all of the conditions, assumptions, and conjectures on which that scientific model is founded were really true? You see, Dr. Yardley, I'm far from convinced evolutionary theorists know, or have any way of proving, whether or not their belief system is capable of getting outside of itself and reflecting anything of the actual nature of reality."

"Objection Your Honor," announced Mr. Mayfield. "My learned colleague is making speeches."

"Yes, sustained," Judge Arnsberger indicated. "Let's move along Mr. Tappin. You'll have time enough for this sort of thing in your closing remarks."

As the counsel for the defense looked over the papers in his hands, he said: "Very well, Your Honor. I apologize to the court for my outburst."

Turning toward the witness, Mr. Tappin asked: "In the Strecker - like, amino acid synthesis scenario that you outlined during direct examination testimony, on what chemical reactants does this kind of synthesis depend?"

"As long as the concentrations of hydrogen cyanide and aldehydes, such as formaldehyde, do not drop too low," pointed out the professor," then researchers believe the Strecker synthesis will be an effective means of converting the aforementioned reactants to amino acids over the course of some 10,000 years."

"What concentration levels," queried Mr. Tappin, "are considered to be minimally necessary for the Strecker synthesis process to be able to proceed?"

"These would be roughly of the order of a 10^{-6} molar solution," Dr. Yardley replied. "This means there should be at least 10^{-6}, or one- millionth, of a mole of solute for each liter of solvent."

"What kind of collective production rates," asked the lawyer, "have been estimated for, say, hydrogen cyanide as a result of ultraviolet radiation, lightning discharges, and shock-synthesis?"

"The figures that I have seen used most frequently," Dr. Yardley answered, "have an upper and lower boundary. These boundaries reflect whether one is talking about a reducing or a relatively neutral atmosphere.

"In the case of a reducing atmosphere such as methane and ammonia, researchers have worked out a production yield of about 100 nanomoles, or 100 billionths of a mole, per square centimeter, per year. This would have resulted in a 3.3×10^{-4} molar concentration of hydrogen cyanide in the Archean era ocean over a period of 10 million years.

"On the other hand, if one were dealing with a relatively neutral atmosphere, the production rate of hydrogen cyanide would have been as much as several orders of magnitude less than 100 nanomoles ... somewhere around 1 nanomole, give or take a few nanomoles ... per square centimeter, per year. Over a ten million-year period, this would have resulted in a 10^{-6} molar concentration of hydrogen cyanide."

"Therefore," Mr. Tappin observed, "the estimated concentration of hydrogen cyanide arising from a relatively neutral atmosphere is right at the minimal limit of what is necessary for the Strecker synthesis to proceed in the Archean era ocean. Is this correct, Dr. Yardley?"

"Yes, that's right," confirmed the professor.

| Origin of Life |

"What assumptions, if any, Dr. Yardley, are made with respect to the conditions in the Archean era ocean in which such a Strecker synthesis is alleged to have taken place?"

"Usually," responded the professor, "researchers assume an ocean pH of either 7 or 8, which is comparable to what we find in the oceans of our present day. Moreover, the temperature of the water is assumed to be about 0 degrees Celsius."

"How do these assumptions, or do these assumptions," inquired the defense counsel, "affect molar concentration estimates for hydrogen cyanide?"

"At a pH of 7 and 0 degrees Celsius, a 3.5×10^{-5} molar concentration of hydrogen cyanide has been calculated for the Archean era ocean. This is based on the reducing-atmosphere production figure of 100 nanomoles per square centimeter, per year."

"Assuming a pH of 8 and, once again, 0 degrees Celsius, one comes up with a 4×10^{-6} molar solution of hydrogen cyanide. This estimate also presupposes the reducing-atmosphere production yield figures cited previously."

"Am I correct in stating," Mr. Tappin asked, "that if one were to use the lower neutral atmosphere production yield rates, rather than the higher, reducing-atmosphere production rates for hydrogen cyanide, then at pH 7 and 0 degrees Celsius, one would have a molar concentration of about 3.5×10^{-7} since the neutral-atmosphere production-yield rates are several orders of magnitude lower than the reducing- atmosphere production rates for hydrogen cyanide?"

"Yes, you would be correct," the professor admitted.

"Professor Yardley," pressed the defense counsel, "is this 3.5×10^{-7} figure for neutral-atmosphere Archean era oceans greater than, or less than, what is minimally needed to be necessary for the Strecker synthesis to proceed in the Archean era ocean?"

"This would be less than what is minimally necessary for the Strecker synthesis to proceed," the professor indicated.

"Moreover, Professor Yardley, " postulated the lawyer, "if one were to assume a pH of 8 and 0 degrees Celsius in the Archean era ocean, as well as presuppose the lesser production-yield figures of a relatively neutral-atmosphere, would one be correct to conclude that this would result in a molar concentration of approximately 4×10^{-8} for hydrogen cyanide?"

"Yes," agreed the professor.

"Is this molar concentration," the lawyer continued, "of 4×10^{-8} for hydrogen cyanide greater than, or less than, the minimal necessary concentration of hydrogen cyanide required for the Strecker synthesis to go forward in the Archean era ocean?"

"Again, this concentration is less than what is minimally required," the professor confirmed.

"Dr. Yardley, what would happen," Mr. Tappin queried, "to hydrogen cyanide concentration figures if one were to raise the temperature of the water to, say, 25 or 50 degrees Celsius, but keep the pH at either 7 or 8?"

The professor hypothesized: "If one were to work on the basis of the reducing-atmosphere production yield figure of 100 nanomoles per square centimeter, per year, then at pH 7 and 25 degrees Celsius, the molar concentration of hydrogen cyanide in the Archean era ocean would be about 2×10^{-8}. In addition, at pH 7 and 50 degrees Celsius, the molar concentration of hydrogen cyanide would be about 3×10^{-9}.

"If, on the other hand ..."

Before the professor could continue, Mr. Tappin interrupted and asked: "Dr. Yardley, don't the figures you are citing indicate that even when one assumes reducing-atmosphere production-yields, which are favorable to the prebiotic evolutionary model, the concentration levels of hydrogen cyanide in the Archean era ocean are insufficient for the Strecker synthesis to proceed?"

"That is correct," the professor admitted.

"Obviously, then," the defense counsel reasoned, "if one were to use the production-yield figures for a neutral-atmosphere, which are several magnitudes of order lower than the reducing -atmosphere production figures, the concentration estimates for hydrogen cyanide in the Archean era ocean would be about 2×10^{-10} and 3×10^{-11}, respectively, for 25-degree Celsius and 50-degree Celsius temperatures at pH 7. Is this right, professor?"

"Yes, it is," Dr. Yardley stated.

"So, these last concentration figures cited," indicated the lawyer, "both for reducing, as well as for neutral-atmosphere production rates of hydrogen cyanide, would be insufficient to sustain Strecker synthesis in the Archean era ocean. This is correct, isn't it?"

"Yes," said the professor.

"Furthermore," Mr. Tappin added, "if one were to work out the concentration figures at pH 8 or pH 9, for either 25 degrees Celsius or 50 degrees Celsius, then, quite irrespective of whether one were working on the assumption of a reducing-atmosphere or the assumption of a neutral-atmosphere, all of the concentration levels for hydrogen cyanide would be far below what is minimally necessary to sustain an amino acid Strecker synthesis in the Archean era ocean. Isn't this the case, Dr. Yardley?"

"Yes, it would be," the professor acknowledged.

"Moreover," the defense counsel continued, "you did previously testify, Professor Yardley, that evolutionary scientists believe the pH of the Archean era ocean was 8, plus or minus 1, did you not?"

"That's right," said the professor.

"Therefore," reasoned the lawyer, "to single out an Archean era ocean with a pH of 7, at 0 degrees Celsius, under conditions of a reducing-atmosphere, is to describe a situation in which everything is stated in terms that are favorable to the idea of a natural account of the origin-of-life from prebiotic conditions. Alternatively, such a way of describing things is to ignore the very real possibilities that the Archean era ocean did not have a pH of 7, or a temperature of 0 degrees Celsius, and might not have existed in conjunction with a reducing-atmosphere.

"All of these other environmental conditions that are possible in the Archean era world would, if true, bring into serious question the plausibility of an evolutionary theory account of the origins-of-life. Would you agree with this, Dr. Yardley?" queried the lawyer.

"If these other possibilities were the case, then, yes, questions of plausibility would begin to arise in relation to such an evolutionary account," admitted the professor.

"What, if any, other assumptions are made concerning the conditions under which the Strecker synthesis is believed to proceed in the Archean era ocean?" Mr. Tappin inquired.

"Well, for one thing," the professor replied, "the Archean ocean is assumed to be comparable in depth and extent to the oceans of today. If the Archean era ocean were shallower or less extensive than current oceans, then, this would serve to increase, somewhat, the concentration figures previously cited. How much this increase of concentration might be, would depend on how much smaller and shallower the Archean oceans were relative to modern day oceans."

"Yet," countered the defense counsel, "couldn't one logically assume that the Archean era ocean was larger, not smaller, than current oceans? After all, the continents had not necessarily established themselves at this period of the Archean era.

"Perhaps, there was more, not less, water during the Archean era, and, therefore, the concentration figures mentioned previously are all inflated somewhat. Isn't this a possibility, Dr. Yardley?"

"Yes, I suppose so," the professor said.

"Isn't it also the case," queried the lawyer "that researchers believe a permanent ice cover formed in the Antarctic only about 20 million years ago, and somewhat more recently in the case of the Arctic region? And, therefore, Professor, might one be correct in assuming that until 20 million years ago, there was quite a bit more water in the Archean era oceans, again diluting the previous concentration figures for hydrogen cyanide?"

"Possibly," Dr. Yardley offered.

"Furthermore, Professor," Mr. Tappin pressed, "doesn't water expand when it is warmer, and if this is correct, isn't there a lot of evidence to indicate that the Archean era atmosphere was sufficiently warm to heat the ocean waters quite a bit above the 0 degrees Celsius temperatures that are being assumed in the Strecker synthesis model, and, therefore, wouldn't this expanded water tend to increase the volume of the solvent, reducing the concentration levels of hydrogen cyanide?"

"Quite possibly," responded the professor.

"Are there any further assumptions," asked the defense counsel, "that frame the conditions under which the Strecker synthesis is believed to have proceeded in the Archean era?"

"There are two more assumptions that I can think of," Dr. Yardley stated. "First, researchers tend to assume all HCN that is produced, by whatever energy pathway, is fully dissolved in the Archean era ocean. Secondly, scientists, generally, assume neither hydrolytic nor thermal degradation will appreciably affect the amount of hydrogen cyanide solute in solution."

"If," postulated the lawyer, "the surface of the Earth were frozen over, as some theorists have proposed in conjunction with the faint early sun paradox, couldn't this affect the amount of hydrogen cyanide that would be able to enter into solution in the Archean era ocean that is alleged to exist below the 300-meter layer of ice?"

"Yes, I guess it could," the professor replied.

"Moreover," Mr. Tappin continued, "if, as some other researchers, alluded to by you, have maintained, there were a layer of between one and ten meters of hydrocarbons floating on top of the ocean, presumably as a result of their non-polar and, therefore, non-soluble nature, then, couldn't this scenario also affect the opportunity of all hydrogen cyanide to enter into solution with the Archean era ocean?"

"I suppose so, yes," indicated the professor.

"In addition," Mr. Tappin pressed, "if we leave aside issues of hydrolytic decomposition, isn't the assumption about the relatively negligible extent of the thermal degradation brought about by hydrothermal vents rather arbitrary and speculative?"

"I believe," the professor offered, "that this assumption about thermal degradation might be based on the roughly ten million years that is required for any given volume of water to circulate throughout the ocean and, presumably pass some given hydrothermal vent. When one compares this period of ten million years to the period of approximately 10,000 years required by the Strecker synthesis, thermal degradation probably would constitute a negligible factor."

"Doesn't this way of thinking," queried the lawyer, "seem to be assuming there is only one hydrothermal vent that is being used as a point of reference for calculating the figure of ten million years necessary for water to completely circulate throughout the ocean? If there were many hydrothermal vents, as might be expected from an early Archean era in which, according to your testimony, some researchers have claimed that the Earth's crust might have been especially vulnerable to such hydrothermal breakthroughs, then the ten million figure that signifies the amount of time required for a given volume of water to circulate through the ocean might be true, but it is irrelevant if many such vents exist at many different points along the bottom of the Archean era ocean. Isn't this so, Dr. Yardley?" asked the defense counsel.

"What you say is a possibility that would have to be taken into consideration, in some way, I suppose," the professor said.

"What about photolysis, Dr. Yardley?" inquired Mr. Tappin. "I noticed you didn't mention this as a possible source of degradation, but wouldn't it have to be factored in, at least with respect to the 200-meter photic zone of the ocean?

"In other words, Dr. Yardley, since all hydrogen cyanide going into solution would have to pass through this photic zone, couldn't photolysis play a major role in affecting the amount of hydrogen cyanide solute available, and, therefore, the molar concentration of this molecule? Moreover, wouldn't this especially be the case given, as you have indicated, that the ultraviolet luminosity of the faint early sun would have been substantially greater during the Archean era?"

"This is a possibility," the professor admitted, "but degradation losses due to things such as ultraviolet light or ionizing radiation are very difficult to measure and, therefore, one has some difficulty in establishing a basis for making estimates in relation to them."

"Whatever the nature of such difficulties, Professor, our ability or inability to measure something really doesn't stop that something from having an effect on us does it?"

"As a matter of fact," stated the professor, "there are interpretations of quantum mechanics that do suggest that reality only comes into being with the act of measurement."

"Dr. Yardley," Mr. Tappin responded, "I believe this is getting more into the realm of philosophy than hard science. However, if you want to begin to grapple with the paradox of how to explain the existence of a prebiotic world prior to the advent of the process of human measurement, I believe you will find evolutionary theory will be in even more difficulty than I, and my client, already believe to be the case."

Looking at the material in his hands, the defense counsel said: "Most of the discussion of the past little while has been about hydrogen cyanide. Very little has been said about formaldehyde, but you previously had stated the Strecker synthesis in the Archean era depends on certain minimal levels of molar concentration being maintained not only for hydrogen cyanide, but for aldehydes such as formaldehyde, as well.

"However," added the lawyer, "in earlier testimony and cross - examination, we established that formaldehyde is not generated during lightning shock-synthesis and also that most of the formaldehyde that might be generated through ultraviolet radiation synthesis is also vulnerable to ultraviolet photolytic degradation. My question, Professor, is this: From whence do the necessary levels of formaldehyde, or other aldehydes, come that are supposed to maintain concentration rates capable of sustaining Strecker synthesis in the Archean era ocean? Even if one could establish requisite production -yield rates, wouldn't all the difficulties that beset the matter of hydrogen cyanide concentration levels also apply to formaldehyde levels of concentration in the Archean era ocean?"

"To the best of my knowledge," Dr. Yardley indicated, "the figures on formaldehyde are less well established than are those for hydrogen cyanide. Nevertheless, I would agree, in general terms, that all of the issues that you have raised in relation to hydrogen cyanide concentration levels would also have to be raised in conjunction with formaldehyde, or other aldehyde, concentration levels in the Archean era ocean."

"Professor Yardley, let's assume," posited the counsel for the defense, "that I was willing to forget all the problems that have been raised with respect to the concentration issue. Do we have any way of knowing what proportion of the amino acids formed in the Archean era ocean through Strecker synthesis would be the twenty varieties of amino acid occurring in living organisms rather than the many other kinds of amino acid that are possible- some of which have been discovered in meteorites?"

"I imagine," answered the professor, "there are individuals with the talent to be able to come up with some kind of thermochemical model that would provide a set of theoretically-driven distribution values for all the different kinds of amino acid that might be possible. However, such a model would be affected by so many variable considerations, conditions and forces, I'm not sure even our current supercomputers could keep track of the problems that would arise in this kind of model.

"One could assume less complex amino acids might tend to be somewhat disproportionately represented in relation to more complex amino acids. On the other hand, a wide array of localized thermodynamic conditions might arise that could run against these sorts of tendencies.

"If temperatures in the ocean were low, say, near 0 degrees Celsius, then one would expect thermal decomposition to be low. However, some amino acids, like alanine and glycine, have far greater stability than do other amino acids, like serine.

"Consequently, stability properties would have to be factored in even if the water temperature were to remain near 0 degrees Celsius, which is unlikely. This is unlikely because within the last twenty to thirty million years there is evidence that bottom water temperatures can vary as much as 10 to 15 degrees as the Earth goes through various climatic transitions.

"What variations in water temperature, top or bottom, might have been taking place across hundreds of millions of years in an Archean era ocean and atmosphere are anybody's guess. Furthermore, how the decomposition tendencies of the twenty amino acids that occur in living organisms would stack up to the decomposition tendencies of all the other amino acids that are possible is another issue that would have to be factored in.

"Then, of course, one would have to work in the decomposing effect that hydrothermal vents and active volcanoes would have on amino acids that had been formed. Since we really don't have any idea of how prevalent either of these processes was during the Archean era, this introduces a further unknown into any prospective model that is being constructed.

"The effects of ultraviolet radiation in the 200-meter photic zone would have to be considered. In addition, once hydrolysis had done its magic and helped amino acids to form, then, the newly -synthesized, more complex amino acids become even more vulnerable to the forces of hydrolysis than is the case for the molecules that reacted together to form them.

"Furthermore, one cannot assume the only sort of synthesis reactions going on in the Archean era ocean are ones that lead to the formation of amino acids. Other, non-amino acid kinds of hydrocarbon are likely to have arisen, and this means there would have been chemical competition for available reactants, with unknown ramifications for the rate and extent of amino acid formation, both in relation to the twenty amino acids that are important to life forms, as well as in relation to the other varieties of amino acid that are not important to life forms on Earth."

"Dr. Yardley, is there," Mr. Tappin inquired, "any reason or mechanism you know of which would have led to the specific selection of the twenty amino acids fundamental to life forms on Earth from among the myriad numbers and kinds of other amino acids that are likely to have arisen in the Archean era ocean through Strecker synthesis?"

"No," the professor answered, "I know of no plausible theory that would explain the selection process that we believe went on during the Archean era. It might well have been a stochastic process, and since we don't know enough about the factors shaping that process, we really cannot do anything but speculate why certain probability distributions might have been thermodynamically and/or kinetically favored over other probability distributions."

"Professor Yardley," continued the defense counsel, "with respect to the amino acids synthesized in the Archean era ocean through the Strecker process, would they have formed a racemic mixture ... that is, a mixture consisting of roughly equal numbers of both left-handed and right-handed optical isomers of the various kinds of amino acid?"

"If our laboratory experiments are any indication, "the professor replied, "then, yes, the Archean era mixture is likely to have been racemic in character. Nevertheless, I previously have mentioned a meteorite found in the Antarctic that contained some exclusively right-handed amino acids, and this discovery does carry some potential implications for what might have occurred in the Archean era ocean."

"Are you aware, Dr. Yardley," asked the lawyer, "of any plausible account that might explain why one might end up with a set of same- handed optical isomers rather than a racemic mixture of amino acids?"

"Over the years," stated the professor, "there have been a number of proposals directed toward this problem of chirality or handedness. The only hypothesis that I have found to be plausible is one proposed back in the 1950s.

"Essentially, this hypothesis assumes that when sunlight passed through the atmosphere of the Archean era, light took on a small degree of polarization. As a result, the polarized ultraviolet component of sunlight during the Archean era might have had a preferential tendency to degrade right-handed optical isomer forms of amino acids, leaving intact the left-handed optical isomer forms that have been observed in the vast majority of Earth organisms."

"Dr. Yardley, don't most of the biologically important carbohydrate molecules tend to exhibit right-handed optical isomer preference?" Mr. Tappin inquired.

"Yes, that's right," the professor indicated.

"So, wouldn't one expect," postulated the lawyer, "that the same polarized ultraviolet component of Archean era sunlight that degraded right-handed amino acid isomers would also degrade right- handed carbohydrate isomers? Consequently, how does one account for the fact one finds right-handed carbohydrate isomers playing fundamental roles in living organisms?"

"This is a problem," Dr. Yardley admitted, "but there might have been other kinds of selection mechanisms at work in addition to the polarized ultraviolet component of Archean era sunlight."

"Does anyone," challenged the defense counsel, "know what these other selection mechanisms were that are assumed to have been operative during the Archean era?"

"Not at this point in time," answered the professor.

"Dr. Yardley," the lawyer hypothesized, "even if one were to accept the polarized-light hypothesis as the reason why left-handed amino acids were selectively favored over right-handed amino acids as far as ultraviolet degradation is concerned, this still leaves at least two problems. First of all, the polarized light assumption doesn't explain why DNA would possess a tendency to call for exclusively left-handed amino acids to be synthesized in the cell. Secondly, one still hasn't explained how the twenty amino acids common to life forms on Earth came to be selectively favored over the other left-handed amino acid optical isomers that would have survived being degraded by slightly polarized ultraviolet radiation. Would you agree with my assessment of the situation, Dr. Yardley?"

"As far as the second problem is concerned," stated the professor, "I would agree no fully satisfactory account presently exists for explaining why the twenty left-handed amino acid isomers were selected over other possible left-handed amino acid isomer candidates. As far as the first problem described by you is concerned, something could be said.

"Selection forces would have favored the DNA and/or RNA system that would have arisen that relied on the optical isomer form of amino acid that was available ... in this case, the left-handed amino acid isomer. If a DNA and/or RNA system would have arisen that depended on the existence of a pool of right-handed amino acid isomers, then given that polarized ultraviolet light had selectively destroyed all, or most, of these kinds of isomer, such a DNA/RNA system would not have been favored by the prevailing conditions of the Archean era world. Prebiotic conditions would have favored the DNA/RNA system that called for, or needed, left-handed amino acid isomers."

"Excuse me, Dr. Yardley, perhaps, I don't understand the situation," said Mr. Tappin. "Although your account or explanation makes sense in the context of having assumed that a left-handed-amino acid-preferring DNA/RNA system already had arisen, your account doesn't really explain how such a left-handed-amino-acid-preferring DNA/RNA system arose in the first place ... does it?"

"No, it doesn't," the professor acknowledged.

"In fact," continued the lawyer, "wouldn't one be justified in arguing that the process of natural selection really is incapable of accounting for change over time except in a post-facto manner? By this, I mean that although natural selection can help explain why certain capabilities, once they arise, might have been selectively favored by existing conditions, nevertheless, natural selection cannot explain how such capabilities arose in the first place, can it, Professor?"

"Well," Dr. Yardley responded, "some theorists do speak in terms of the idea of 'evolutionary pressure'. In other words, they believe the collective character of any given set of conditions might, in a sense, generate a certain amount of pressure to induce the sort of changes that would be favorably selected by those conditions."

"How does this process of inducement work?" Mr. Tappin asked. "How does the physical/chemical world induce a given system to change both its structural character, as well as its way of operating, so that the system adopts a structure and set of processes that would be selectively favored by the prevailing conditions of that physical and chemical world?"

"It's a very complicated issue," replied the professor. "There is a great deal of work going on with the science of complexity, as well as chaos theory and the theory of dissipative structures that is directed toward trying to answer questions like this."

"Has anyone," inquired the lawyer, "come up with a model in any of these disciplines that has been accepted by the scientific community as a plausible account of how prevailing physical and chemical circumstances induce a system to generate structural and dynamic changes that are, capable of taking advantage of precisely the conditions that prevail in the world at a given time?"

"Not yet," responded the professor.

"Then, Dr. Yardley, would one be doing injustice to the available evidence," Mr. Tappin pressed, "if one were to say, at least at this point in time, that the notion of evolutionary pressure is a totally unproven hypothesis however convenient and desirable an idea it might be for evolutionary theory?"

"No, I would have to say" the professor admitted, "that no injustice would be done to the available evidence."

"Consequently," summarized the defense counsel, "currently, there really is no plausible, generally-accepted explanation of how or why DNA or RNA systems arose that showed a preference for left-handed amino acid isomers as well as right-handed carbohydrate isomers. Would you agree with this statement, Dr. Yardley?"

"Yes, at the present time, what you have said is the case," agreed the professor.

"Mr. Tappin, I'm going to exercise some discretion and intervene at this juncture," Judge Arnsberger indicated, "to propose that court be adjourned for lunch. Court will reconvene again at 2:00 p. m. this afternoon."

Monkeying Around With The Containment Blues

As Mr. Tappin rose from behind the defense table he took the material being handed to him by a member of his team. He started to walk toward Dr. Yardley, stopped and retraced his steps.

He leaned over and whispered something in the ear of his colleague. When he received an affirmative response, he straightened up.

On his way back to the area near the witness stand, he was busy inspecting the new batch of material. He continued to do so for a further ten seconds, or so, after stopping in front of the witness stand.

Finally, he said: "In your direct examination testimony you referred to an experiment by Fox in which urea [$CO(NH_2)_2$] and malic acid ($C_4H_6O_5$) were heated to 150 degrees Celsius under conditions free from water ... that is, which were anhydrous in nature. You indicated this experiment resulted in the synthesis of aspartic acid.

"In a further experiment, also performed by Fox, you talked about a recipe for generating polymers or bonded chains of amino acid. In this recipe, if one cooked the amino acid glutamic acid in an oil bath for one hour at 170 degrees Celsius, and, then, blended in a variety of other amino acids and cooked the whole mixture for a further three hours at the same 170 degrees Celsius, then one could produce a chain of amino acids consisting of up to a hundred units.

"In variations on this experiment, phosphoric acid was added, and the variables of time and temperature were played around with during different runs of the same experiment. This resulted in an increase in the amounts of neutral and basic amino acids that could be incorporated into the polymer chain of amino acids.

"You also described another experiment in which sunlight was passed through a solution of paraformaldehyde ($CH_2O)_3$, ammonia and ferric chloride. After a certain amount of time, this arrangement brought about the synthesis of the amino acids serine and asparagine.

"During direct examination testimony, you talked, as well, about an experiment by Oró in which hydrogen cyanide, ammonia and water were combined to produce, over a period of time, a number of different amino acids. In addition, a certain amount of the purine, nucleic base, adenine, showed up as a product in this experiment.

"You also discussed how when the foregoing set-up was altered somewhat, other kinds of molecules could be synthesized. For instance, if one combined cyanogen (C_2N_2) and cyanoacetylene (HC_3N) with hydrogen cyanide (HCN), then one could obtain other nucleic bases such as uracil, cytosine, guanine and thymine.

"Finally, in another experiment performed by Oró you outlined, first, how he took some fatty acids, one of the fundamental building blocks of many important lipids, and, then, how he dried these fatty acids in the presence of phosphate and glycerol. In this manner, simple phospholipids, that are fundamental components of membranes in living organisms, were synthesized.

"I must admit," Mr. Tappin indicated, "on the one hand, I find all of this experimental ingenuity quite impressive. On the other hand, I also find such ingenuity potentially troublesome.

"More specifically, Dr. Yardley, different ingredients are taken from here and there and mixed together in certain ways, for particular lengths of time, under specified conditions of temperature, acidity, and so on. In other words, Professor, the requirements for these experiments are all different from one another, involving and depending on different conditions, reactants and treatment.

"Presumably, these experiments are intended to simulate prebiotic conditions and demonstrate how purely natural processes could lead to the synthesis of organic compounds that have potentially important implications for origin-of-life issues. However, just as was true in Miller's original origin-of-life, I'm having trouble understanding how these experiments simulate actual prebiotic conditions and processes.

"For example, Dr. Yardley, do we have any way of telling how prevalent such materials as urea, malic acid, paraformaldehyde, ferric chloride, cyanoacetylene, cyanogen, fatty acids, phosphate, and glycerol would have been in the Archean era?"

"We believe," answered the professor, "that most of the compounds you listed would have been available, some more so than others, during the Archean era. Most of these compounds are extremely simple in structural formula, and we believe they would have been formed relatively easily through natural chemical processes going on during that period of time."

"Dr. Yardley, correct me if I am wrong, but fatty acids are hardly simple hydrocarbons." Referring to the sheets in his hand, he added: "Let's see ... palmitic acid, which is one of the most abundant saturated fatty acids, has a formula of $CH_3(CH_2)_{14}COOH$. Oleic acid, which is one of the most common unsaturated fatty acids, has a formula of $CH_3(CH_2)_7CH{:}CH(CH_2)_7\text{-}COOH$."

"Wouldn't you agree, Professor, that oleic acid and palmitic acid have considerably more complexity than hydrogen cyanide (HCN), ammonia (NH_3) and methane (CH_4)?"

"Yes," Dr. Yardley acknowledged.

"I believe," suggested the defense counsel, "that in your direct examination testimony you said the Fischer-Tropsch reaction was involved in bringing about some of the steps necessary for the formation of fatty acids. Is my recall on this matter accurate, Dr. Yardley?"

"Yes, it is," stated the professor.

"Would you please review once more for the members of the jury, Professor, the general nature of the Fischer-Tropsch process," requested Mr. Tappin.

"One takes a gaseous form of carbon, like carbon monoxide (CO)," the professor explained, "together with water vapor, and, then one passes these over a hot iron-powder catalyst, at temperatures between 180 and 300 degrees Celsius and under anywhere from one to fifty atmospheres of pressure."

"Will one have fatty acids at the end of this process?" asked the lawyer.

"No," replied the professor. "After the foregoing procedure has been run, one must find a way to oxidize the hydrocarbon chains that have been generated by means of the Fischer-Tropsch mechanism."

"In your opinion, Dr. Yardley," asked the defense counsel, "how likely would a naturally occurring counterpart to the Fischer-Tropsch reaction be?"

"The fairest thing I can say" the professor suggested, "is that a naturally occurring counterpart to the Fischer-Tropsch reaction is extremely unlikely but not entirely inconceivable. When one adds to this the requirement of a further oxidation step, one is really pushing the envelope of credibility to the outer limits."

"In the Oró experiment mentioned earlier," indicated the lawyer, "from which phospholipids were synthesized -- two further ingredients were needed in addition to fatty acids ... namely, glycerol and phosphate. How available were these molecules likely to have been in prebiotic times?"

"This is hard to say. The structural formula for glycerol is $C_3H_8O_3$ and is normally formed from the decomposition of natural fats by means of an alkali compound or superheated steam.

"There might have been some series of natural chemical reactions during prebiotic times that was capable of synthesizing glycerol. The structural character of this compound is not so complex that the act of assuming the existence of such a hydrocarbon during the Archean era strains credibility.

"A phosphate, on the other hand, is produced by combining an alcohol group with any one of three phosphoric acids. For instance, orthophosphoric acid, which is quite stable, has the formula H_3PO_4.

"Phosphorus, one of the main ingredients of phosphates and phosphoric acids, is a fairly rare non-metallic element. Even at the best of times there are only trace amounts of phosphorus to be found in seawater, and the presence of phosphorus in the Earth's crust is quite limited relative to elements such as magnesium, iron, calcium, potassium, sodium and silicon.

"Phosphates are very rare in nature, although human beings are quite adept at dumping huge quantities of these compounds into the environment. However, as far as prebiotic times are concerned, there would be no obvious, plentiful source of phosphates, and, therefore, phosphates would not have been readily available to support, in a rigorous fashion, any reaction requiring them during the Archean era.

"This does not mean there were no phosphates in prebiotic times. It merely means their relative scarcity would have placed constraints on where, when, and how frequently phosphate-dependent reactions could have proceeded."

"Dr. Yardley, could one fairly say," inquired the lawyer, "that the plausible likelihood of not only producing, but, as well, bringing together, fatty acids, glycerol and phosphates in order to synthesize phospholipid compounds under prebiotic, Archean era conditions is seriously in question?"

"Yes," the professor agreed, "I think one would not be unfair if one were to characterize the situation in this fashion. This doesn't necessarily mean the whole thing is completely impossible, but at this point in time, in the

| Origin of Life |

149

light of what is known, many researchers can't imagine any series of plausible steps during prebiotic times that, one, would have led to the formation of the individual reactants involved in phospholipid synthesis, or, two, would have resulted in these ingredients coming together to make such a reaction possible."

"Therefore," reasoned the defense counsel, "to call Oró's phospholipid synthesis experiment a simulation that accurately reflects what went on under the Archean era's prebiotic conditions is really, potentially, quite misleading. Would you agree with this, Dr. Yardley?"

"Let's just say" the professor offered, "the indicated potential to be misleading is present, and one cannot treat the natural, prebiotic synthesis of glycerol, phosphates, fatty acids or phospholipids as foregone conclusions. At best, the issue lends itself to being highly contentious and argumentative."

"Dr. Yardley, let's return to the Fox polymerization experiment for a moment," Mr. Tappin suggested. "A recipe was used in that experiment that called for a variety of amino acids to be thrown into a mixing bowl of sorts. Subsequently, these ingredients were heated for some 3-4 hours in an oil bath at 170 degrees Celsius.

"In your direct examination testimony, Professor, you indicated many researchers believe the exposed surface of a sandy beach, or a mineral bed, or a strip of solidified lava, where temperatures might have reached up to 100 degrees Celsius, might have served as a crucible for certain condensation reactions during the Archean era. In another portion of your testimony, you spoke about hydrothermal vents in which the temperatures were in the vicinity of 350 degrees Celsius, but these took place under water, not in oil.

"You didn't specifically speak about the conditions around volcanoes in your testimony, Professor. Yet, since neither of the previously-mentioned possibilities really matches the required conditions of the Fox experiment, can one assume that, perhaps, the area in and around certain volcanoes is the only other candidate that, conceivably, might fit into the kind of scenario that Fox's proteinoid experiment is purporting to simulate?"

"Volcanic areas," the professor said, "seem to be the only possibility that comes readily to mind."

"Would you agree, Dr. Yardley," inquired the lawyer, "that finding a place in volcanic areas that provided an oil bath of precisely 170 degrees Celsius for just 3-4 hours would be ... let's be kind here ... a tricky project?"

"Yes," responded the professor, "I guess one might not find many places capable of meeting these precise conditions, but this is not the same thing as saying that these sorts of conditions couldn't or didn't, exist."

"Professor Yardley, in your testimony concerning the Fox experiment, you mentioned, I believe," recalled the defense counsel, "that not all of the bonds that linked together the amino acid monomers or units were peptide in character ... that only some of these bonds were peptide in character. Is this correct?"

"Yes," the professor replied.

"In living organisms on Earth, peptide bonds," the lawyer stipulated, "occur between the amino and carboxyl groups of neighboring amino acids, binding them together to form proteins. Isn't this so, Dr. Yardley?"

"That's right," the professor confirmed.

"Therefore," concluded the defense counsel, "the amino acid polymers or chains in Fox's experiment are not really proteins because they are not what we find in living organisms. Presumably, for precisely this reason, the polymers in Fox's experiment are called proteinoids and not proteins. Is this a fair way of putting things, Dr. Yardley?"

"I guess so," admitted the professor.

"Did any of these proteinoids exhibit substantial enzymatic characteristics?" inquired Mr. Tappin.

"Not really," the professor stated. "On the other hand, there might not be anything that prevents proteinoids from playing the other major role of proteins involving the morphology ... that is, the form and structure ... of organisms.

"Conceivably, a variety of ribozymes ... in other words, polymers of RNA with enzymatic properties ... might have served as the early enzymes of the protocell. Proteinoids could have filled the function of helping to give form to these protocells or to various organelles such as ribosomes or mitochondria, within the protocell."

"Is it not the case, Dr. Yardley," queried the lawyer, "that the bonds, whether peptide or otherwise, formed during condensation reactions in which water is removed from neighboring monomeric amino acids and, therefore, are called anhydride bonds ... isn't it the case these anhydride bonds are quite labile and, relatively speaking, easily broken."

"Yes, under certain conditions, this is true," the professor acknowledged.

"Would you agree, Dr. Yardley," asked Mr. Tappin, "that volcanic areas in which temperatures are 170 degrees Celsius, or higher, for prolonged periods of time, might be considered to have met the requirements alluded to by you through your use of the qualification: 'under certain condition', with respect to the labile nature of peptide bonds among amino acids?"

"Yes," admitted the professor.

"Are we not encountering here," wondered the lawyer, "yet another instance in which, under certain conditions, energy might be coupled to chemical reactants for short periods and in specific ways, to forge more complex arrangements of hydrocarbons, but when, under other circumstances, these same forms of energy can quickly turn the tables on the products of such reactions and, as a result, undo what these energy forms previously had helped to bring about?"

"Yes, this is a possibility," the professor agreed.

"In describing the Fox proteinoid polymerization experiment, Dr. Yardley, you said that, by playing around with the time and temperature variables, Fox was able to incorporate more neutral and basic amino acids into the proteinoid polymers synthesized through condensation reactions. Is this right?" Mr. Tappin inquired.

"Yes," affirmed the professor.

"In effect, Dr. Yardley, doesn't this mean," pressed the lawyer, "that if we are to consider the Fox experiment to be a simulation of Archean era conditions, then, not only must we assume there were specialized pockets in which amino acids could gather together in an oil bath for 3-4 hours at precisely 170 degrees Celsius, but there were also other pockets in these volcanic areas in which amino acids could be bathed in oil for slightly less, or slightly more, than 3-4 hours, at temperatures that were somewhat higher, or somewhat lower, than 170 degrees Celsius so that proteinoids

| Origin of Life |

152

with greater numbers of neutral and basic amino acids could be incorporated into these polymer chains?"

"Yes," stated the professor. "We believe the entire Archean era world was a prebiotic version of a modern laboratory in which there were many different kinds of evolutionary niche being explored. In these various pockets, millions, if not billions, of different sorts of experiment were being run across the several hundred million years required for protocells or primitive organisms to emerge.

"At this time, I should add," Dr. Yardley indicated, "there have been experiments in which polypeptide polymers have been observed to form in the absence of water when mixtures of amino acids were incubated at a temperature of 65 degrees Celsius for a period of 40 days. So, one doesn't have to be tied to the 170-degree Celsius figure of the early proteinoid experiments."

"Wouldn't you agree, Professor Yardley," suggested the lawyer, "that finding a little corner of the Archean era world that will allow one to incubate a mixture of amino acids at 65 degrees Celsius for precisely forty days, twenty-four hours a day, no more or no less, is really only a variation on the problem that is being discussed?"

"I suppose so," the professor responded, "but this latter experiment does introduce a broader spectrum of possibilities into the picture."

"Let us assume, for the moment," Mr. Tappin proposed, "that, as a result of some of the points brought out previously under cross-examination, the amino acids used in the simulation experiments of Fox, or this more recent 65 degree/40-day experiment, were not forthcoming from Strecker synthesis in the Archean era ocean. Given this assumption, how would these amino acids find their way into the mixing bowl pockets or crucibles of the different volcanic areas?"

"As a number of experiments have indicated," the professor stated, "there are a variety of alternative pathways to amino acid formation other than Strecker synthesis."

"Would," inquired the lawyer, "urea $[CO(NH_2)_2]$, malic acid $(C_4H_6O_5)$ and paraformaldehyde $[(CH_2O)_3]$ ... which are just three of the reactants used in laboratory experiments in order to help synthesize a few, specific amino acids ... would these compounds have been readily available in the Archean era?"

"How readily the various compounds cited by you would have been available might be an issue of some debate," the professor offered, "but we believe there was a reasonably good chance such compounds would have been synthesized under various conditions during the Archean era."

"Would this last answer remain the same, Dr. Yardley," queried the defense counsel, "if one were to raise the same kind of question in conjunction with cyanogen (C_2N_2) and cyanoacetylene (HC_3N) that, together with hydrogen cyanide (HCN) have been used in laboratory experiments to synthesize nucleic bases such as uracil, cytosine, guanine and thymine?"

"Yes, my last answer would remain substantially the same," the professor stated.

"Dr. Yardley, do any of the alternative pathways to which we have alluded produce all of the amino acids?" Mr. Tappin asked. "In other words, in accordance with what has been established previously through testimony and cross-examination, aren't these pathways frequently quite specific in terms of the reactants, temperatures, and conditions that are necessary to generate certain kinds of amino acid?"

"This is often the case, yes," the professor confirmed, "but not always. Some methods have produced a number of different amino acids by varying the experimental conditions slightly, although, as you have indicated, no one method has generated all of the amino acids."

"If no one method has generated all of the amino acids," hypothesized the lawyer, "could one reasonably argue there might have been some physical distance that might have separated these pathways from one another since these alternative pathways often presuppose different precursor reactants, different temperatures, and so on?"

"I guess one could argue in this fashion," the professor acknowledged, "but I don't think one can assume great distances were necessarily involved. Many of these reactions could have happened in, and around, the same volcanic areas."

"Alternatively, Professor," the defense counsel pointed out, "one cannot necessarily assume relatively great distances were not involved either, can one?"

"No, one can't," Dr. Yardley conceded.

"If," Mr. Tappin postulated, "one assumes the Strecker synthesis process, followed by tidal movement to intertidal zones, was not the primary means of delivering amino acids to places where condensation reactions could take place, is there a secondary or backup account of how amino acids generated from different chemical pathways and under different conditions would have come together in Fox's prebiotic mixing bowl?"

"I suppose," the professor replied, "one would have to speak in terms of chance, random processes in order to account for how these kinds of events might be possible."

"Is this an explanation, Dr. Yardley, or an assumption?" asked Mr. Tappin.

"In other words, if one has no reliable baseline from which to construct distribution models that permit one to demonstrate how a series of unrelated and complex events might reasonably be anticipated to come together, what exactly is being explained? Isn't one merely assuming something has happened in a particular way and labeling that assumption with the name of 'chance events'?"

"Not entirely," the professor asserted. "If one were to take a large enough group of monkeys and put them together with a sufficiently large set of typewriters, then, mathematically, one could predict, with a fair amount of confidence, that, sooner or later, one of the monkeys would type a perfect copy of, say, Hamlet."

"What about," the lawyer wondered," *The Glass Bead Game* by Hesse or, since we seem to be dealing with science fiction here, something by Isaac Asimov?"

"Objection Your Honor," Mr. Mayfield stated. "Learned counsel is being rather frivolous in his questioning at this point."

"Your Honor," Mr. Tappin countered, "since I have encountered the witness' argument before, under other circumstances, and since the example of Hamlet was often the work cited in this kind of argument, I was curious as to whether these monkeys were stuck in some sort of creative rut and were unable to write anything else."

"As was true in the case of the proverbial cat with the same propensity," Judge Arnsberger replied, "this sort of curiosity is not likely to have a long lifetime in my courtroom. You've made your point, Mr. Tappin, let's move on. The prosecution's objection is overruled."

"Your Honor," asked Dr. Yardley, "may I be permitted to answer the question?"

"Certainly," the judge responded, "but you are under no obligation to do so."

"I understand, Your Honor," acknowledged the professor, "but, nevertheless, I would like to address the question."

Turning back toward the defense counsel, the professor said: "In theory, there is no limit on the nature of the books that could be produced by these monkeys. So, Hesse's work or the *Foundation* series by Asimov, both would be possibilities, or, if you like, you can even throw in some Raymond Chandler."

"Dr. Yardley," inquired Mr. Tappin, "wouldn't one be able to predict, with considerably more confidence, and based on empirical evidence rather than on a mathematics rooted in contentious and unprovable assumptions, that, sooner or later, all of the typewriters would be destroyed, all the paper would have been used up, and the monkeys would have been dead long before so much as the thought, let alone the typed reality, of even a coherent paragraph of any kind would have occurred to these monkeys, whether considered collectively or individually?"

"Objection, Your Honor," Mr. Mayfield interjected.

Before Judge Arnsberger could speak, Mr. Tappin announced: "I'll withdraw the question, Your Honor.

"Let's assume," postulated the defense attorney, "the mathematical theory to which you are alluding is true. How large would the set of typewriters and group of monkeys have to be in order for a copy of, for example, Hamlet, to get written by one of the monkeys, and how long would all of this take?"

"We are dealing here with the mathematics of the infinite," stated the professor. "If one had an infinite number of typewriters, monkeys and paper, then, at some point, Hamlet would emerge.

"The interesting possibility in all of this is that, given such starting assumptions, Hamlet might very well get written within a finite length of time since there is no way to pin down where in the infinite series of events the desired copy of Hamlet would be forthcoming. The book might appear after 10,000 years or 10,000,000 million years or 100,000,000 million years, and even though these numbers are very large, they are finite, and, more importantly, they are reminiscent of the sort of time considerations involved in origin-of-life issues."

"This mathematical theory, Professor, seems to be assuming," Mr. Tappin suggested, "that in any given single striking action, all keys of the typewriter have an equal opportunity of being struck by any given monkey, with no single striking trial having any influence on the striking actions that precede or follow it. In other words, each striking action of the moment is entirely independent from all other striking actions, whether performed by the same monkey or by other monkeys. Would you agree with this Dr. Yardley?"

"Yes, I suppose so," the professor agreed.

"Your mathematical theory appears to be assuming, as well," the defense counsel continued, "that every possible sequence of key-striking events, eventually, will be represented by the activities of the monkeys. Furthermore, since the sequence of key-striking events that makes up or constitutes the work of Hamlet would be one such set of sequential key-striking events, then, one has opened the door for the possibility that at least one of the sets of independent key-striking events will give expression to a sequence that matches Hamlet word for word.

"Is the foregoing a fair way of describing the situation?" the lawyer asked.

"I believe" replied the professor, "the reasoning of the theory runs, more or less, along the lines you have indicated."

"Has anyone tested this mathematical theory empirically?" inquired the defense counsel.

"I'm not quite sure what you mean," the professor said.

"Has anyone, for instance," Mr. Tappin specified, "attempted to determine whether or not the assumption of independence with respect to key-striking action is warranted in the context of the activities of real rather than theoretical monkeys? Or, has anyone tried to discover whether all sets of sequential key-striking activity are equally represented or whether some sets are over-represented or underrepresented?"

"No, I don't think anyone has tried any of what you are suggesting," the professor responded.

"Has anyone attempted to discover," queried Mr. Tappin, "whether monkeys would continue to type from hour to hour, day to day, week to week, and month to month as a demonstration of their capacity, in principle, to be able to produce any kind of effort that would be comparable in length to the work of *Hamlet?*"

"Not really," answered the professor.

"Dr. Yardley, were there an infinite number of molecules on the surface, or in the atmosphere, of the Archean era Earth?" asked the lawyer. "Or, were there an infinite number of chemical reactions that went on during the Archean era? Or, was there an infinite amount of energy available to run those reactions?"

"No, of course not," the professor said.

"Then," Mr. Tappin proposed, "what might, or might not, happen in a universe of infinite monkeys, typewriters and paper, really doesn't constitute an appropriate way of modeling objects, processes and events that are finite in nature, does it?"

"Perhaps not," admitted the professor, "but the basic principle is, nonetheless, suggestive. Given large numbers of even finite chemical events, then, certain kinds of events might become more likely over the long run, although these same events might appear to be very unlikely in the short run."

"Wouldn't the projected likelihood of such events depend on the nature of those events?" the defense lawyer inquired. "Wouldn't one have to be able to provide some good reason why, in the long run, one might reasonably expect events with a specific character to occur that one would not anticipate would take place in the short run?

"More specifically, Professor, do we really have any reasons aside from, or independent of, the vague notion of chance events, which would permit us to suppose that in the long run we reasonably can expect a bunch of amino acids that are generated through different pathways and under different circumstances to all end up in the same place at the same time? Moreover, if we don't have anything independent of the notion of chance, random events with which to work, then, aren't we back where we started ... namely, isn't this a matter of assumption rather than a matter of scientific proof or demonstration?"

"Your Honor," stated Mr. Mayfield, "I must object. This question already has been asked of, and answered by, the witness. We are going over the same ground."

"Overruled," Judge Arnsberger proclaimed. "I'm going to allow the question."

Dr. Yardley was silent for about ten seconds or so. When he spoke, he said: "Stochastic models provide a way of setting parameters without presupposing any particular kind of metaphysics or ontology. These models offer an opportunity to explore and analyze what does happen against frameworks of expectation and anticipation based on the general properties and characteristics of natural phenomena.

"To say that some given event ... such as the coming together, at some point in time and space, of a variety of amino acids generated through separate pathways and conditions ... has a finite, although small, possibility of occurring is doing nothing more than to recognize that real events often are capable of reflecting different aspects of our stochastic models. The perfect bridge hand or throwing 'x' number of consecutive passes at the gaming tables, or winning a lottery against huge odds, and so on, constitute, as far as our stochastic models are concerned, very rare events, but they do happen.

"In fact, the more runs of any given activity that take place, the greater, in general, will be the likelihood of seeing theoretical possibilities being realized or manifested in actual circumstances that one would not expect, on the basis of one's stochastic model, to occur with any degree of frequency. Although the chemical events taking place during prebiotic times might not have been infinite in number, nevertheless, the number of such reactions over the course of four to eight hundred million years is incredibly high.

"Given such large numbers, one might expect, at some point, that certain kinds of improbable events have a chance of taking place. I don't consider such an improbable event an assumption, however unlikely it might be, since its possible occurrence is rooted in a complex stochastic modeling process that acknowledges these kinds of event to be conceivable and capable of taking place in finite, real time."

"When you say, Dr. Yardley, that something is 'capable of taking place in finite, real time', are you saying," Mr. Tappin asked, "that this something must take place or necessarily will take place, or, are you merely saying the event in question could take place under the right circumstances?"

"I'm saying," the professor indicated, "that such an event could take place under the right circumstances and that such circumstances can be assigned some small, but finite, probability of actually occurring."

"What is the nature of this process of assigning some small but finite probability?" the lawyer asked.

"The nature of the assignment process would be shaped by the character of one's stochastic model," replied the professor. "Different models might assign different kinds of probability to this kind of situation."

"Are any of these assignment procedures based on empirical data?" inquired Mr. Tappin.

"Yes, they could be," the professor stated. "It depends on what one is talking about."

"How about," proposed the lawyer, "the coming-together of twenty left-handed-amino-acid-isomers of the sort that are observed to occur in Earth organisms?"

"Well," began the professor, "one would have to figure out how many different kinds of amino acids could have been synthesized under prebiotic conditions. One, then, might, or might not, multiply that number by two, depending on whether one believed ultraviolet light had been polarized slightly in its passage through the Earth's atmosphere and, as a result, had a tendency to decompose right-handed amino acid isomers.

"One also would have to try to work out frequency distribution tables for the different kinds of amino acids, including the 20 in which you are interested. These frequency distribution tables would depend on such things as production efficiency yields and energy efficiency yields for the various stages of amino acid formation that we discussed earlier in the context of the Strecker synthesis process.

"In addition, these frequency distribution tables would have to reflect, in some way, how many amino acids came from extraterrestrial sources. On the other hand, one would have to factor in losses due to pyrolysis, hydrolysis, photolysis, absorption by various clay materials, and so on.

"When one took all of these factors into consideration, one would be in a position to calculate theoretical values about what proportion of the total set of amino acids in existence at any given time were represented by the 20 left-handed amino acids you mentioned. This would provide some sort of stochastic baseline to apply to the real world and from which one's expectations concerning these possibilities would arise.

"Professor Yardley, has anyone worked all this out?" Mr. Tappin asked.

"Models have been developed that take various combinations of these factors into consideration," Dr. Yardley answered. "However, to the best of my knowledge, no one has taken all of these factors into consideration. At this time, we simply don't have the software, models, and computers capable of handling the complex dynamics that result from the interaction of all these variables."

"Would one, therefore, Professor, be incorrect in saying there is no complete model of what went on during the Archean era as far as amino acid formation is concerned?" inquired the lawyer?"

"No, this would not be incorrect," acknowledged the professor. "On the other hand, the very essence of science is a constant process of improving, revising, updating, modifying, and, sometimes, rejecting the models that are being constructed.

"Science doesn't purport to have the final answers," added the professor. "It is a work-in-progress, and, as such, it attempts to do the best it can with the materials that are available to it.

"As new material, techniques, ideas, and methods have become available the evolutionary model has been able to improve upon its past performance. The revisions and modifications that have come through this process of gradual, conceptual evolution have created a more rigorous model, but we continue to seek to improve it."

"Given what you have just said," hypothesized the defense counsel, "would one be fair, Dr. Yardley, if one were to say the following? If one does not wish to call the assignment of a probability concerning the likelihood of a bunch of amino acids coming together in the general vicinity of some volcano an 'assumption', then, could one fairly say the stochastic model responsible for assigning probabilities in this case stands in need of considerable revision?"

"I don't have a problem with this way of stating things," the professor indicated.

"Dr. Yardley, in all of our discussions up to this point, concerning the different kinds of experiments that have been conducted in relation to origin-of life issues, is it not the case that the various experiments were run with purified compounds under conditions in which there was no chemical competition going on among different kinds of compounds to determine which compounds would form covalent bonds with which compounds?" Mr. Tappin inquired.

"I would say so, yes," the professor replied.

"Would you agree, then, Dr. Yardley," asked the lawyer, "that one might have difficulty understanding how simple condensation cycles of heating and drying might bring about a very selective synthesis of pure polymers, such as proteins, DNA, and RNA -- with the right kinds of bonds, optical activity, and monomer composition -- from amongst the highly complex mixture of hydrocarbons that might have been available as reactants in the Archean era world?"

"I would agree there is a challenge here for evolutionary theory," admitted the professor, "because there still are quite a few things we don't, yet, understand. I would not agree this challenge necessarily constitutes an insurmountable barrier to our being able to understand these issues eventually.

"Our knowledge base," pointed out the professor, "is developing exponentially. Furthermore, the interim periods required for our knowledge to double are becoming increasingly shorter.

"Phenomena that were inexplicable a few years ago are now being understood. To acknowledge the existence of a problem or challenge is to participate in the natural order of things in the world of science."

"Professor, consider the following hypothetical situation," requested Mr. Tappin. "Suppose there were a relatively dilute, Archean era, ocean solution of phosphates, carbohydrates, pyrimidines, purines, fatty acids, amino acids, and various kinds of other simpler hydrocarbons.

"Let us further suppose, Dr. Yardley," added the lawyer, "that some of this seawater solution finds its way, via tides and the wind, to some intertidal zonal, or lava, surface. What is likely to happen once this dilute solution starts to get heated from the sun and/or volcanic- related activity?"

The professor considered the hypothetical situation briefly and began to speak. "Probably, as evaporation proceeded, then, at some point, sodium chloride crystals would form. Bivalent cations, or positively charged ions and radicals, would interact with organic anions, or negatively charged hydrocarbon groups. Finally, there would be a very large number, and variety, of covalent bonds that would join together different functional groups in virtually every conceivable combination."

"Would you expect," the counsel for the defense inquired, "that such a mixture of ions and covalent bonds would organize itself thermodynamically into a working protocell?"

"If you are asking me," posited the professor, "whether I would expect something interesting to happen in the single exposed lava surface or intertidal puddle that is being examined hypothetically, then I would have to say no, I would not expect such a mixture to organize itself into a working protocell. However, if you were asking me about my expectations in relation to billions of such exposed surfaces and/or intertidal puddles, then I would have to say, yes, I would begin to feel confident in my expectations that at least one of these prebiotic crucibles would be capable of thermodynamically and kinetically organizing itself into something very interesting as far as the origin-of-life issue is concerned."

"Do we have anything," the lawyer queried, "beside your rising level of felt confidence in such expectations that is likely to persuade us there is something inevitable or necessary about the possibility that, at some time and at some place, there must be a protocell that must emerge from the prebiotic mists? After all, Professor, if you are relying on billions and billions of exposed lava surfaces and intertidal puddles to give rise to at least one interesting protocell or near -protocell, then, the prima facie odds against this sort of event happening are billions and billions to one, wouldn't you say?"

"As I indicated earlier," the professor replied, "the more opportunities there are for experimentation with different combinations of possibility, then, the greater is the probability that one of these sets of combinations will possess and exhibit the sort of characteristics and properties in which one is interested as far as origin-of-life issues are concerned."

"Dr. Yardley, you seem to be assuming," the lawyer suggested, "that all of these billions and billions of prebiotic crucibles will necessarily be exploring all conceivable possibilities. However, what guarantee do we have that these mini-laboratories, even if they are in the hundreds of billions and trillions, will be sufficient to explore all the possible combinations available to the molecules in the dilute solutions that have washed up on various exposed surfaces or into some intertidal puddle?"

"Naturally," responded the professor, "there can be no such guarantee."

"Moreover," Mr. Tappin continued without pausing, "what guarantee do we have that even if, on the basis of thermodynamic theory, a given combination is considered possible that, therefore, from a kinetic perspective, every such thermodynamically conceivable combination will actually occur."

"Again," said the professor, "there can be no guarantee in such matters."

"Or," the lawyer added, "how do we know there won't be a tendency in such mini-laboratories, due to various thermodynamic or kinetic considerations, to repeat, again and again, some finite, but large, set of prebiotic experiments at the expense of other possibilities, and, in the process, consume a great deal of the resources of materials, space, energy and time that are available?"

"All I can say," remarked the professor "is that, in general terms, you have raised a number of valid issues that need to be addressed. However, the fact these problems have been raised doesn't preclude the possibility of discovering either answers to your challenges or of finding ways that open up the possibility of side-stepping or circumventing these problems in some way."

"At the present time, Dr. Yardley, does evolutionary biology have any remotely satisfying answer for the problems being raised here -- yes or no?" specified the lawyer.

"I would have to say no," answered the professor.

Returning to the defense table, Mr. Tappin went through the, by now, well-established ritual of exchanging new material for used material with his colleague. As the lawyer turned toward the witness, he began speaking.

"Professor Yardley, during an earlier part of cross-examination, we talked about the difficulty of plausibly accounting for the generation, and bringing together, of compounds such as fatty acids, phosphates and glycerol in order to try to synthesize phospholipids, one of the primary components of many kinds of cell membrane. Before proceeding to talk about cell membranes in a little more detail, there is one further point that I would like to address.

"I believe phosphatidic acids are the simplest class of phospholipids," the lawyer said. "Is this correct?"

"Yes," replied the professor.

"Moreover," Mr. Tappin added, is it also the case that derivatives of phosphatidic acid, such as lecithin, tend to exist in cells primarily in an optical isomeric form that is in a left-handed rather than in a right- handed isomeric configuration?"

"That's right," the professor confirmed.

"Consequently, Professor," Mr. Tappin concluded, once again, evolutionary theory is confronted with the problem of having to come up with an explanation for how such a preference arose with respect to optical isomers, just as in the case of proteins, as well as of ribonucleic acids. Would you agree with this assessment of the situation?"

"Yes, I would," acknowledged the professor.

"If, Dr. Yardley, as presently seems to be the case based on present knowledge, there is no readily apparent, natural pathway by which to generate phospholipids, how do evolutionary biologists propose to account for the development of cell membranes?" inquired the lawyer.

"There are a number of different possibilities," the professor stated. "In my earlier testimony, I touched on a number of these, including carbonaceous chondrites, proteinoid micro spheres and transitional liposome-like structures."

"Would you expand a little, Dr. Yardley, on the possible role of carbonaceous chondrites with respect to cell membrane formation?" the lawyer requested.

"There are several ways to look at the findings vis-à-vis carbonaceous chondrites," the professor began. "One of these ways involves the discovery of amphiphilic compounds, and the other possibility deals with the hydrocarbons that are found in some of these meteorites.

"Amphiphilic compounds," explained the professor, "have both: hydrophilic, or water-loving, as well as hydrophobic, or water-hating, components. These compounds have been observed to spontaneously form membranous-like boundary structures when placed in an aqueous environment.

"When placed in water, the hydrophobic parts of these compounds tend to curl up in order to minimize contact with water. In the process of curling up, a vesicle or protected, interior space is created within which various kinds of chemical reaction might take place under the right circumstance."

"Dr. Yardley, before you continue on," Mr. Tappin interjected, "I would be interested to know if tests have been conducted to determine if these amphiphilic compounds exhibited any phospholipid -like properties?"

"Samples of these compounds were studied by means of an electron microscope," responded the professor. "One of the purposes of this analysis was to determine if a membranous structure was present in these compounds.

"These studies did detect the presence of a membranous structure approximately 10 nanometers, or 10 billionths of a meter, in thickness. This is consistent with the upper boundary size of the cell membranes of many organisms.

"In addition to the electron microscope studies, tests were performed in order to examine the ability of these membranous structures to encapsulate polar solutes, or water-soluble molecules, in a manner that was the same as, or similar to, cellular membranes in living organisms. A dye was used in this study and the researchers found that the amphiphilic material from carbonaceous chondrites had the ability to encapsulate polar solutes with approximately one-tenth of one percent of the encapsulation efficiency of the phospholipids found in living organisms."

"In other words, Dr. Yardley, although these extraterrestrial compounds could form membrane-like structures with about the same thickness as the cell membranes of living organisms, they were almost nothing like phospholipids in this, presumably, important area of being able to encapsulate polar solutes. Is this correct?" the defense counsel asked.

"Essentially, yes," the professor responded.

"You also mentioned, Professor, the hydrocarbon-related possibility associated with the carbonaceous chondrites," the lawyer said. "What exactly does this involve?"

"Around 1970," the professor pointed out, "several researchers studied seven carbonaceous chondrite meteorites. They discovered chains of hydrocarbons consisting of between 10 and 23 carbon atoms - a finding that was consistent with what also had been observed in the Murchison meteorite.

"This is comparable, in some respects, to the 12 to 20 carbon atoms contained in fatty acids, one of the main components of the lipids found in the phospholipids that make up most cell membranes. In the absence of any plausible natural prebiotic method of synthesizing fatty acids, such chains might have served as a source for the type of hydrocarbons that make up fatty acids in lipids and, therefore, cell membranes."

"Dr. Yardley, isn't it the case," asked the defense counsel, "that fatty acids contain chains of hydrocarbons consisting of even numbers of carbon atoms?"

"That's right," the professor acknowledged."

"Therefore," said the lawyer, "not only are some of the hydrocarbon chains, ranging in length from 10 to 23 carbon atoms, which are found in the meteorites, both too short or too long, relative to those hydrocarbon chains that range in length from 12 to 20 carbon atoms that are found in fatty acids, but if the meteorite hydrocarbon chains contain odd numbers of carbon atoms, then, this would be another dissimilarity between the meteorite hydrocarbons and fatty acid hydrocarbons. Is this correct, Dr. Yardley?"

"Yes," the professor replied.

"In effect, if one tried to view these differences in the best possible light," stated the lawyer, "one would have to assume that, somehow, carbon atoms either would have to be added to, or removed from, many of the hydrocarbon chains found in the meteorites. Would you agree with this, Dr. Yardley?"

"This seems reasonable," indicated the professor.

"Furthermore," Mr. Tappin continued, "isn't it the case that the hydrocarbon chains found in the meteorites would have to be oxidized before those hydrocarbon chains, with the right lengths of even numbered carbon atoms, could be considered to be fatty acids?"

"Most probably," the professor answered.

"In addition," Mr. Tappin pressed, "even if one were to concede that fatty acids might arise in the prebiotic Archean era world in this extraterrestrial fashion, one still would have to find a way to bring these fatty acids together with phosphates and glycerol, under the right conditions, in order to synthesize phospholipids. And, given that phosphates, in particular, are likely to be extremely rare compounds in the Archean era, then, Dr. Yardley, wouldn't one have to consider this whole sequence of events to be very, very improbable?" the lawyer asked.

"I imagine this would be the case," affirmed the professor.

"Finally, in the light of previously established testimony," the lawyer stipulated, "one cannot assume meteorites would represent a very substantial source of these kinds of hydrocarbon chains, nor can one assume these hydrocarbon chains necessarily would survive post- impact, prolonged exposure to ultraviolet photolysis or, perhaps, even heat, in the form of, possibly, relatively high surface temperatures or volcanic activity. Isn't this so, Professor?"

"Yes," Dr. Yardley agreed, "one cannot assume these sorts of thing to be automatic or given."

"As far as the possible role of proteinoids is concerned in relation to membrane functioning," the defense counsel queried, "is there any evidence, Dr. Yardley, that proteinoids have the necessary properties to form active transport systems, or establish ion pump mechanisms, or to provide transmembrane channel ways, as proteins do in the membrane complexes of living organisms?"

"At the present time, I believe there is little, if any, evidence to suggest proteinoids have the kinds of capability to which you are referring," replied the professor. "Nevertheless, the absence of evidence in the few laboratory experiments that have been performed to date does not preclude the possibility that during the Archean era, proteinoids with some of these sorts of functional capacity might have been synthesized naturally."

"Is there any evidence, Dr. Yardley, that the proteinoids have the necessary sort of sequential arrangements of hydrophobic and hydrophilic amino acids that, upon folding into their tertiary or folded structure by means of thermodynamic forces, will enable their folded hydrophobic portions to be located in the interior portions of the phospholipid bilayer and, consequently, match up with the hydrophobic hydrocarbons of the lipid molecules, as is the case in the transmembrane proteins of living organisms?"

"At this time, I know of no such evidence," Dr. Yardley admitted.

"Previously, Professor," pointed out Mr. Tappin, "you talked about liposomes. You described them as small vesicles made up of lipid bilayers that might have served as a transitional membrane-like structure.

"To talk about liposomes, of course, is assuming that the issue of lipid formation had been resolved in the Archean era. Would you agree with this, Dr. Yardley?"

"Yes," the professor said.

"While elaborating on the structural character of liposomes," said the lawyer, "you spoke about properties such as the ability to reverse breakage of the bilayer by spontaneously resealing any gaps that occur as a result of, say, mechanical agitation or shaking. In addition, Dr. Yardley, you mentioned liposome properties such as being able to trap solutes that might happen to be nearby when these vesicles are dried, as well as liposome qualities of growth, division and multiplication ... all of which are reminiscent of what goes on in living organisms.

"Growth, division and multiplication, Professor, all suggest having access to a supply, regular or irregular, of lipid molecules. Consequently, wouldn't you agree these properties of growth, and so on, all presuppose that additional lipid molecules will be available which, in turn, means that, once again, the question of lipid availability in the Archean era would have to be addressed?"

"Yes," iterated the professor.

"Do liposomes control their own growth, division and multiplication, Dr. Yardley, or is this alleged growth, division and multiplication something that sometimes occurs to liposomes as a result of external forces impinging on the liposomes, or as a result of, say, osmotic lysis ... that is, the rupture of the liposome due to an inward diffusion of salt and water in the process of establishing an equilibrium between internal and external environments of a liposome?

"If," postulated the professor, "you are asking me whether the liposome can be said to be alive in some sense, then, clearly, the liposome cannot be described as being alive, nor does it control its growth, division and multiplication in the same sense that a biological organism actively controls these processes. On the other hand, the capacities of a membrane structure to reverse breakage, expand in size, and be able to participate in processes of division and multiplication, are fundamental stepping stones on the road toward becoming part of the life phenomenon."

"If my understanding on the matter is correct, Dr. Yardley, living cells are, within certain limits, able to maintain an internal electrical or ionic potential that is different from the surrounding environment. In fact, some people have suggested that the ability of a bounded, or membrane-enclosed, system to maintain this kind of differentiated energetic relationship with the environment is one of the most recognizable attributes of a living organism. Would you agree with this way of characterizing the situation?"

"Yes," the professor confirmed.

"Are liposomes capable of maintaining this kind of differentiated energetic relationship with the environment?" asked the lawyer.

"No," Dr. Yardley stated. "As I indicated before, liposomes are not living organisms."

"What happens, Professor, if some sort of potential difference arises between the internal and external regions of a liposome?"

"Lipid structures," Dr. Yardley stated, "tend to show considerable permeability to water, as well as a small amount of permeability to positively and negatively charged ions of low molecular weight, although these ions diffuse across the membrane at a rate that is about one billion times slower than is the case for water molecules. Therefore, whenever there is disequilibrium between the inner and outer environments of the liposome, osmotic diffusion occurs, and this tends to eliminate the disequilibrium.

"If these potential differences are slight, then, equilibrium might be re-established with no appreciable effect on the bilayer structure of the liposome's membrane. If the potential differences are great, say, in favor of the external environment relative to the internal environment of the liposome, then, the liposome will swell with the osmotic diffusion of ions and water into the vesicle's interior and, eventually, might undergo lysis or rupture."

"Dr. Yardley, what would a liposome-like structure need in order to get around this osmotic problem?" Mr. Tappin inquired.

"One would need," replied the professor, "either some kind of rigid wall capable of resisting the stresses of lysis, or one would need a system capable, as required by circumstances, of pumping ions in an out of the interior of the structure, or one would need some combination of rigid walls and an ion pump."

"When you say 'rigid wall', this, presumably, refers to things like cellulose in plants," queried the lawyer.

"Yes," the professor answered. "However, fungi, bacteria, and algae have evolved a variety of rigid structures besides cellulose to handle the problem of osmotic lysis.

"Some of these alternative strategies involve combinations of polysaccharide molecules that are different from cellulose. Other strategies for creating rigidity in membrane walls also have arisen, involving, for example, silica, lime and chitin ... an amorphous polysaccharide that is intermediate between proteins and carbohydrates ... in conjunction with, say, various carbohydrate matrices."

"Would one be fair, Dr. Yardley, if one were to say that phospholipids require the presence of particular kinds of protein in order to have ion pumping capabilities so that even if one were to assume phospholipids were laying around, so to speak, in the Archean era, nonetheless, the mere presence of phospholipids, in and of themselves, would not solve the osmosis problem?"

"Yes, that's right," indicated the professor.

"Therefore," the lawyer said, "attaching just any old kind of proteinoids, or even proteins for that matter, to phospholipids will not necessarily establish an ion-pumping capability, unless these proteinoids or proteins have the right kind of sequential, structural, and tertiary folding properties that are suited to transporting particular kinds of ions into and out of the membrane-enclosed structure. Is this right, Dr. Yardley?"

"I would say so," the professor replied.

"Presumably," Mr. Tappin hypothesized, "various kinds of proteinoids or proteins would be necessary to handle the transport or pumping of different kinds of ions such as sodium, magnesium, potassium, calcium, and so on. Would you agree with this, Professor?"

"Yes," Dr. Yardley said.

"This capacity of a membrane system to actively participate in accepting some things while excluding others is referred to as 'selective permeability', isn't it?" the lawyer asked.

"That's correct," acknowledged the professor.

"Besides ions, Dr. Yardley, what other kinds of capability," the defense counsel inquired, "would need to be actively included or excluded if a membrane-enclosed structure were to possess the full range of functional characteristics exhibited by the membrane systems of living organisms?"

"Organisms would need some means of actively transporting nutrients into the interior of the cell," the professor stated. "Simultaneously, organisms would need a means of not only getting rid of toxic materials that might be accumulating as a result of the catabolic and anabolic ... that is, respectively, the tearing down and synthesizing ... processes going on in the cell in relation to such nutrients, but there would have to be some way for this active transport system to be able to selectively differentiate toxic materials from metabolites being used in the cell."

"Presumably," the lawyer reasoned, "different kinds of transport mechanisms across the membrane and/or channel ways through the membrane would be needed in order to bring different kinds of nutrient into the cell, as well as carry various sorts of toxic material out of the cell. Would you agree with this, Dr. Yardley?"

"Yes, I would," affirmed the professor.

Mr. Tappin asked: "Why couldn't nutrients and toxic substances just enter and leave the cell by means of osmotic diffusion, in the same way water and low molecular weight ions do in liposomes?"

Dr. Yardley explained: "The phospholipid molecules that form the bilayer structure characteristic of membranes, constitute a hydrophobic permeability barrier to all hydrophilic, or water loving, materials, as well as to high molecular weight ions. Passive diffusion, or osmosis, will not carry these kinds of compounds across the permeability barrier formed by the phospholipid bilayer, and, therefore, active forms of transport must be used, or channel ways must be provided that will allow unimpeded passage through the hydrophobic interior of the bilayer membrane structure."

"What would happen," Mr. Tappin hypothesized, "if the nutrients transported across the membrane were not coordinated with the organism's ability to catabolically tear down, and then anabolically build up necessary molecules using these kinds of nutrient?"

"The organism would starve to death," responded the professor.

"In other words," continued the lawyer, "being able to actively transport nutrients across the membrane's permeability barrier is not enough. These nutrients must be of the right kind, and, therefore, would one be right in supposing, Dr. Yardley, that this particular transport mechanism must be able to preferentially select those nutrients that will be of use to the organism?"

"Yes, I suppose this would be the case," said the professor.

"Isn't it true," queried the lawyer, " that modern bacterial organisms tend to divide about every twenty minutes or so, and, consequently, they need to transport enough phosphates, of one sort or another, across their membranes, in the interval between divisions, to be able to double the supply of these molecules that are crucial to the process of synthesizing the increased amount of ribonucleic acids required for cell division?"

"Yes, that is right," the professor indicated.

"Moreover, isn't it the case, Dr. Yardley," asked the defense counsel, "that because phosphates tend to be ionized, a specialized carrier enzyme is necessary for the capturing and transporting of phosphates across the permeability barrier formed by the cell membrane of these bacteria?"

"Yes," agreed the professor.

"Consequently," the lawyer concluded, "to look after processes of selective permeability ... such as ion-pumping, nutrient or toxic transport, along with phosphate acquisition and carrier requirements ... one needs a variety of proteinoids or proteins with specialized amino acid sequences to give one the structural characteristics, hydrophobic or hydrophilic properties, and tertiary folding patterns that meet such a diverse array of cellular needs. Therefore, not just any kind of proteinoid or protein structure will serve such purposes, is that right Dr. Yardley?"

"As far as we know, this is the way things work," the professor confirmed.

"Do membranes provide functions other than the ones already mentioned, Dr. Yardley ... that is, other than ion-pumping and active-transporting, mechanisms of one kind or another?" the lawyer inquired.

"The ability to maintain a differentiated energetic potential between the interior and exterior environments of the cell," pointed out the professor, "establishes an ion gradient. This gradient represents a mother lode of energy that can be mined in various ways to serve a number of cell functions, including coupled transport of nutrients that already has been touched on to some extent and the production of compounds like adenosine triphosphate (ATP) that becomes a mobile means of supplying energy to chemical processes going on throughout the cell.

"For many years," the professor added, "scientists have known that if one heats and then dries a phosphate solution, an anhydride bond forms between pairs of phosphate molecules. This anhydride bond is able to store the energy that is released by the heating and drying process.

"The pair of phosphate molecules that are bonded by the anhydride bond are known as pyrophosphate molecules. Adenosine triphosphate, along with a number of other kinds of phosphate compounds such as creatine phosphate and phosphoenolpyruvate, contain pyrophosphate bonds that are capable of storing energy.

"Essentially, in the case of the potential electrical difference that has been established across the membrane's permeability barrier, the ion gradient becomes the source for generating the energy that is stored in the pyrophosphate bonds of ATP rather than through the energy that is released by the aforementioned laboratory method of heating and drying of a phosphate solution."

"So," the defense counsel proposed, "in order to have a protocell begin to self-assemble, not only do we need to come up with a solution of phosphates in the Archean era, we also need to find a way to generate, at a minimum, the anhydride bonds of pyrophosphates so that we have a means of storing energy generated by the ion gradient associated with the cell membrane ... providing, of course, we can manage to find a way to get these pyrophosphate bonds into the interior of the bounded environment formed by a complex of phospholipids and proteinoids. Does the foregoing scenario cover, in broad terms, this aspect of the evolutionary perspective, Dr. Yardley?"

"In broad terms, yes," replied the professor.

"Stripped down to its bare essentials, Dr. Yardley, would one be right to say," Mr. Tappin asked, "that the mining of the energy contained in the ion gradient being maintained by the potential electrical difference between the interior and the exterior of the cell ... would one be correct if one were to describe this mining process as the rolling, so to speak, of electrons and/or protons down the gradient in order to gain the energy generated by the downhill movement of these charged particles along the ion or proton gradient?"

"This is, more or less, accurate," acknowledged the professor "although, as you indicated, your description is obviously an extremely simplified version of what actually occurs in the energy producing reactions that take place along the ion gradient established by the potential electrical difference across the cell membrane."

"Dr. Yardley, in living organisms, isn't this process of electron or proton translocation along the electrical gradient that extends across the membrane, handled by specific enzymes or proteins?" inquired the lawyer.

"That's correct," the professor said.

"Therefore, in addition to the specialized proteinoids or proteins needed for the pumping of ions, as well as the transport of compounds such as phosphates, nutrients and toxic materials, one also needs specialized proteinoids or proteins capable of translocating electrons or protons across the membrane's ion gradient in order to be able to transfer the energy potential of that gradient to pyrophosphate bonds in compounds such as adenosine triphosphate. Is this the case, Dr. Yardley?"

"Yes, it is," affirmed the professor.

"Given," postulated the lawyer, "a phospholipid bilayer that is impenetrable to all ionic molecules except ones of very low molecular weight, and given that many proteins contain not only ionic side chains but hydrophilic components, how does evolutionary theory account for the process that would allow proteins to become embedded in a permeability barrier that, due to its hydrophobic character, one might assume would be resistant to such a process?"

"We believe," Dr. Yardley stated, "there is some sort of thermodynamic driving force that would allow the proteins and the phospholipids to overcome the repulsive forces acting between the two kinds of molecule. This chemical antagonism is inherently unstable.

"Conceivably, this condition of disequilibrium could be resolved if there were some, as yet undiscovered, thermodynamic process that allowed the energy of the system to be re-distributed in a more stable arrangement. Presumably, the embedding action might take place during this process involving the thermodynamically driven ... and, therefore, spontaneous ... redistribution of the energy toward a more stable ground state."

"You did say, Professor, this thermodynamic mechanism for the insertion of proteins into phospholipid bilayers was both theoretical and, as of yet, undiscovered, is this right?" queried the lawyer.

"Yes, I did," Dr. Yardley admitted. "However, the fact proteins are found embedded in phospholipid bilayers in living organisms, despite the inherent chemical antagonisms that are involved and the fact we have not seen any evidence of a kinetic or non -thermodynamic mechanism to account for this state of affairs, then, the thermodynamic hypothesis outlined above, although theoretical and unproven, is not as speculative and arbitrary as you might think."

"Has anyone," Mr. Tappin asked, "come up with a non-protein related way of mining the energy of the ion gradient that exists in conjunction with the cell membrane?"

"Over the years, a lot of different theories have been proposed in this regard," the professor remarked. "These usually concern variation on themes involving some kind of electron tunneling, ion migration, or proton transfer.

"So far, however, there doesn't appear to be a plausible way of making these mechanisms capable of working in any consistent, reliable fashion, or capable of generating the levels of energy that would be required to maintain membrane functioning, not to mention many other cellular processes. In addition, even if one could come up with a viable, non-protein-related mechanism for mining energy from the membrane's ion gradient, there is no way of either storing the energy once it reaches the interior of the cell, nor is there any way of transferring the charge in order to chemically activate other molecules involved in cell processes, since, as far as is known, both the storage of charge as well as the charge-transfer processes are effected by proteins, although the energy storage compound, itself, is often some kind of a nucleotide rather than a protein."

"Dr. Yardley, would you agree," inquired Mr. Tappin, "that even if one could come up with a plausible prebiotic theory for, one, the migration of charge across the permeability barrier of the membrane, two, the storage of charge, and, three, the transfer of charge, all of which we will assume are capable of operating quite independently of proteins, wouldn't one still be faced with the problem of having to explain how the non-protein system evolved to produce the protein- based system that now helps govern charge-migration, charge-storage and charge-transfer in the biological organisms with which we are presently familiar?"

"Yes," acknowledged the professor. "I don't see how one could avoid having to address this problem under such circumstances.

"In fact, in my opinion, this is precisely the sort of difficulty that emerges in relation to theories of the origin-of-life that focus on the possible role of clay minerals. The proponents of these theories talk about the capacity of clay surfaces to carry out some of the functions important to life ... such as exhibiting a few catalytic properties that can help bring about certain stages in the polymerization of some of the nucleotides in nucleic acids, as well as some peptide chaining; or, providing a surface on which concentration reactions can take place; or, offering a means to compartmentalize and organize different metabolic pathway; as well as having the potential to store, and replicate, certain kinds of information on crystalline patterns, somewhat reminiscent of genetic system. However, in point of fact, even if one were to ignore all the problems and rather severe limitations that surround such capabilities in mineral clays, like kaolin and montmorillonite, nonetheless, these theorists have no way of explaining how life, as we understand it, came into being.

"In effect, they avoid the real problems surrounding origin-of-life issues by trying to define life in another, very limited and superficial way. As a result, they tend to multiply the theoretical problems because not only must they account for the rise of such clay mineral photocells, these theorists also must come up with a plausible theory of transition that accounts for the genetic takeover of these clay mineral systems by protocells that are not based on clay minerals ... unless, of course, such clay mineral protocells are not part of our evolutionary lineage, in which case, whether the theory is right or wrong, it really has nothing to do with life as we understand it.

"Above and beyond the foregoing, there is a further problem concerning the viability of a clay mineral hypothesis for the origin of life. Many clays -- including kaolin -- tend to be extremely rare in pre-Cambrian sediments.

"This fact does not constitute a fatal blow to these kinds of hypothesis. On the other hand, such a fact does tend to lessen the chances of such a hypothesis being correct.

"Quite frequently, one will find various kinds of inorganic conjectures thrown into the picture in an attempt to augment or complement the clay mineral origin-of-life hypothesis. For instance, relatively recently there was a conjecture by a European theorist that is based on the manner in which iron sulfides, like pyrite, contain free energy when the iron becomes reduced to a ferrous state.

"Using such an observation as a launching pad, this theorist postulated that, possibly, if one could find a way of coupling this free energy to possible reactants in a protocell-like environment, then, an important component in the formation of one or more primitive metabolic pathways would have been established. When one added that this kind of energy source might tend to be found in close contact with, say, clay mineral surfaces that, among other things, were capable of bringing about concentration reactions, such a conjecture became quite attractive to some people.

"However," Dr. Yardley concluded, "no plausible, dependable means has been found for accounting how the charge-transfer, or coupling, process will take place in conjunction with potential chemical reactants in a protocell-like environment. Therefore, the iron sulfides conjecture remains nothing but an unrealized conjecture.

"Similarly, some people have proposed that when the various components of nucleotides ... ribose, phosphate, and a nucleic base of one kind or another ... are adsorbed onto the surface of some clay mineral, then, perhaps, the specific character of the mineral might have brought these components together in particular orientations. Unfortunately, for this kind of proposal, none of the minerals that have been tested to date have exhibited the requisite specificity to be able to generate nucleotides with the sort of structural character that are observed in living organisms."

"In conjunction with the previous discussion of membrane activity and functions," Mr. Tappin specified, "isn't it the case that various classes of pigments might be involved with the processes of photosynthesis that take place in, and about, the thylakoid membranes in photosynthetic bacteria and blue-green algae, as well as the chloroplasts of plants?"

"That's right," answered Dr. Yardley.

"What role does porphyrin play in all of this?" the defense lawyer asked.

"Porphyrins," explained the professor, "are one of a group of pigments that are widely distributed among different kinds of organisms. They are derived from a porphin molecule that is a ring structure made up of four pyrrole nuclei (C4H4NH) linked together by carbon atoms.

"The nitrogen atom in porphins often tends to form very strong and stable bonds with metallic ions such as magnesium or iron. This kind of bonded group is referred to as a chelate.

"Chlorophyll, which is present in all photosynthetic organisms, consists of a porphin group with a magnesium ion at its center. In addition, different kinds of chlorophyll have various kinds of side chains attached to them.

"Generally speaking, pigments are divided into two broad classes known as accessory and principle pigments. Accessory pigments tend to gather light energy and pass it onto the principle pigment that, for the most part, is either chlorophyll 'A' or one of the forms of chlorophyll occurring in certain bacteria.

"There are, however, other classes of non-chlorophyll pigments such as carotenoid and phycobilin. These other classes of pigments tend to have accessory, rather than principle, roles in photosynthetic systems."

"Professor Yardley, to the best of your knowledge," inquired the lawyer, "is there any plausible prebiotic pathway of synthesis that might give rise to the Porphyrins that are at the heart of the chlorophyll contained in all photosynthetic organisms?"

"None is known at the present time," replied the professor. "Nonetheless, as I indicated in previous testimony, on occasion, pigment-like molecules have been found in the organic residue of some carbonaceous chondrites."

"Even if," Mr. Tappin postulated, "we were to assume these pigment-like molecules had a full capacity to accept and transfer light energy, and even if we were to assume these extraterrestrial pigments were in plentiful supply and did not get degraded through photolysis and so on, and even if one were to assume that, somehow, these pigment-like molecules were to find their way into a protocell system, wouldn't one still be faced with the problems of explaining how porphin-containing chlorophyll came into existence and how these pigment-like molecules became coordinated with chlorophyll molecules in various kinds of photosynthetic systems?"

"Yes," the professor conceded, "one still would be left with having to account for such things."

"Furthermore, Dr. Yardley, in the photosynthetic systems with which we currently are familiar, doesn't the transfer of energy charge from accessory to principle pigments take place by means of an electron transport system made up of a series of protein enzymes, and, therefore, even if one were to accept the idea of an extraterrestrial pigment-like molecule playing a role in the formation of early photocells, wouldn't one still need to account for the rise of the requisite support system of enzymes that had the ability to serve as a specific transport mechanism in relation to the movement of electrons to their final acceptor destination in the protocell?"

"Yes," the professor acknowledged, "these sorts of phenomena would remain as problems to be explained ... but even in the assumptions that you have cited there are also chemosynthetic autotrophic organisms that derive their carbon and energy in a quite different manner from photosynthetic autotrophic organisms. Conceivably, these chemosynthetic autotrophs, and not photosynthetic autotrophs, were the first photocells to exhibit the properties of life."

"If I understand what you are saying, Dr. Yardley, wouldn't evolutionary biology now have two problems to solve rather than one?" suggested the defense counselor. "The origin of two different kinds of autotrophs would have to be accounted for ... one that is chemosynthetic in nature and one that is photosynthetic in nature. Is this the case?"

"It is," stated the professor," unless one of the two systems were the prototype from which the other eventually was derived through an evolutionary process."

"If this were the case, wouldn't one still be faced with two problems?" Mr. Tappin challenged. The first problem would be to provide a plausible explanation for either photosynthetic or chemosynthetic autotrophs, depending on which one an individual considered to have arisen initially. The second problem would be to provide a plausible explanation for the sort of transitional steps that would have permitted a very different kind of autotrophic system to be derived from the first autotrophic system. Isn't this the situation, Professor, with which evolutionary biology would be, and is, faced?"

"Yes, I suppose it would be, and I suppose it is," Dr. Yardley responded.

"Mr. Tappin," stated Judge Arnsberger, "once more, I must interrupt your cross-examination. The dinner hour is at hand, and I feel we all could use a break from these deliberations.

"Please remember, all of my previous instructions to the jury remain in effect. These court proceedings will be adjourned until 7:30 p.m. this evening."

The Science of Presumption Can Be a Beautiful Thing

"Dr. Yardley," stated Mr. Tappin, "you have testified that ribose is a 5-carbon monosaccharide or pentose sugar monomer. In addition, you said this sugar, along with phosphates and nucleic bases, are fundamental building blocks of nucleic acids, and nucleic acids are the carriers of genetic information.

"How do evolutionary theorists account for the synthesis of ribose sugars in the prebiotic Archean era?" asked the defense counsel."

"Many researchers feel," the professor replied, "that a process known as the formose reaction might have been the most plausible means for synthesizing a variety of sugars including ribose. Essentially, this involves a base-catalyzed condensation reaction of formaldehyde."

"Leaving aside for the moment," said the lawyer, "the previously established point concerning the possible, relative unavailability of formaldehyde in a prebiotic environment due to, among other things, ultraviolet photolysis, would you describe in a little more detail the nature of the formose reaction."

"If," began the professor, "one takes a strong alkali agent such as thallium hydroxide or lead hydroxide and treats formaldehyde with one or the other of these agents, one can generate a variety of sugars. On the other hand, one also can use agents like alumina ... that is, aluminum oxide (Al_2O_2), as well as calcium carbonate or barium hydroxide.

"Following an induction period ... which might last for many hours and in which products such as glycolaldehyde, glyceraldehyde and dihydroxyacetone are formed ... a variety of sugars are synthesized. These include tetroses, pentoses and hexoses, or, respectively, 4 -, 5-, and 6-carbon sugars.

"The formose reaction is autocatalytic in nature which means that once the induction period is over, the reaction proceeds to completion rather quickly. In addition, if the reaction is stopped at the appropriate stage, yields of up to 50% of some of the higher sugars are possible."

"Dr. Yardley, since, presumably, there was no one around in prebiotic times to stop the formose reaction at the appropriate stage, can one reasonably assume that the yields would have been considerably less than the 50 percent figure you have cited?" Mr. Tappin inquired.

"Yes, I guess so," indicated the professor. "On the other hand, there could have been forces active in the prebiotic environment that might have disrupted the reaction before it went to completion."

"I won't pursue this Archean era version of a mugging by unknown assailants," the defense counsel remarked, "but I would like to pursue the issue of the alkali agents that might be used in the formose reaction. How common would, respectively, thallium, lead, and barium hydroxide have been during the Archean era?"

"This is relatively difficult to say," the professor responded. "Perhaps the most accurate thing I can say is these hydroxides probably would have been far less plentiful than either aluminum oxide, which is very common in the silicates that make up a large portion of the Earth's crust, or calcium carbonate - that is, limestone, which also would have been quite plentiful in the prebiotic period."

"Is there," Mr. Tappin asked, "only one kind of pentose sugar -- such as ribose -- which is synthesized during the formose reaction?"

"No," replied the professor. There are a number of pentoses that are formed during this reaction, and each of these pentose sugars are produced in varying amounts.

"For example, in addition to ribose, one also will find xylose, lyxose, and arabinose. These other pentoses involve various kinds of inversion of one or more of the hydroxyl groups of ribose."

"What proportion of all the different kinds of tetrose, pentose, and hexose sugars formed during the formose reaction," queried the defense counsel "are the ribose variety of sugar?"

"Ribose forms a very small portion of the overall yield of sugars," the professor stated.

"Do the other pentose sugars beside ribose get synthesized in amounts that are comparable to, if not more than, the ribose yields?" inquired the lawyer.

"Yes, they do," answered the professor.

"What sorts of concentration levels of formaldehyde are minimally necessary for the formose reaction to proceed?" Mr. Tappin wondered.

"As far as we know," the professor stipulated, "the formose reaction does not seem to proceed if the solute level of formaldehyde falls much below one-hundredth of a mole per liter of solution."

"Given," postulated the lawyer, "what has been said before about the possible scarcity of formaldehyde in the Archean era ... and, perhaps, even in the best of circumstances ... aren't expectations for the existence of such high solute concentrations of formaldehyde during prebiotic times rather inflated and optimistic?"

"Yes, realization of these levels of formaldehyde concentration during the Archean era could be a significant obstacle to the formation of ribose," confirmed the professor.

"Dr. Yardley, how stable are sugars in aqueous solution?"

"Not very," the professor replied, "especially if the pH value is above 7. Under these circumstances, sugars tend to be degraded over a period of time that is not much longer than what is required to synthesize such molecules."

"Previously, Professor, you stated that evolutionary researchers usually consider the pH of the Archean era ocean to have been 8 -- plus or minus one. Consequently, would you agree, Dr. Yardley, the pH of the Archean era ocean had a very good chance of exceeding a pH of 7 and, therefore, readily could have led to the destruction of whatever small amounts of ribose were synthesized almost as quickly as these molecules were formed."

"Yes, there could have been a very good chance this happened if the pH of the Archean era ocean was much above 7," affirmed the professor.

"Other than the issue of isomers with different-handed optical activity, does ribose come in more than one form?" the defense counsel inquired.

"Yes, it does," the professor replied. "There are three forms in all. "In addition to a form known as ribopyranose," he explained, "there are two ringed forms of ribose. These are referred to as alpha and beta-ribofuranose."

"Do all three of these forms of ribose appear in the nucleic acids that occur in living organisms?" asked Mr. Tappin.

"No," stated the professor. "The only form of ribose that occurs in living organisms is beta-ribofuranose."

"Nucleosides," stated the lawyer, "are one step removed from a full-fledged nucleic acid due to the absence of a phosphate group, and nucleosides consist of bonding together one of the five nucleic bases with a beta-ribofuranose. Have I got this right?"

"Yes," the professor indicated.

"Could other sugars, such as some of the non-ribose pentoses, bond with the five nucleic bases?" inquired the defense counsel.

"Yes," Dr. Yardley confirmed.

"Presumably," surmised the lawyer, "all three forms of ribose also could form bonds with the nucleic bases. Is this correct?"

"Yes, that is right," said the professor.

"Consequently," Mr. Tappin concluded, "any one of a number of pentose sugars, or different forms of ribose, or optical isomers could bond with the nucleic bases and form one species, or another, of a nucleoside. Yet, only one of the nucleosides, amongst this mixture of possible nucleosides, has any functional value in living organisms. Would you agree this is the case, Professor?"

"I would," Dr. Yardley acknowledged.

"How," the lawyer queried, "did the one nucleoside that would have functional value once living organisms arose come to be selected from the multiplicity of very similar choices available in the Archean era environment?"

"We are not sure," Dr. Yardley admitted. "Obviously, whatever the mechanism of selection, the beta-ribofuranose nucleoside had selective value."

"What exactly do you mean, Professor, by the notion of selective value?" asked the defense counsel.

"The beta-ribofuranose nucleoside worked," the professor responded. "It fit in with the rest of the protocell system and, presumably, played a fundamental role in forming a self-sustaining, and self-perpetuating, system."

"Wouldn't you say this is a matter of twenty-twenty hindsight?" challenged Mr. Tappin. "Before one reached the stage of establishing even a primitive protocell, one would have to assume the beta-ribofuranose nucleoside is being selected.

"One cannot use the functioning of a system," argued the lawyer, "that has not yet been established as the reason for why such a molecule is being selected. So, why is this particular molecule, among all the other possibilities, being selected for, prior to the existence of a working protocell?"

"One can only assume," the professor stated, "that this particular nucleoside must have satisfied certain thermodynamic and kinetic contingencies that existed during the Archean era."

"Are the identities of these contingencies to be kept anonymous at this time, Professor?"

"I'm afraid so," acknowledged the professor. "I should point out, however, that Albert Eschenmoser, of the Swiss Federal Institute of Technology, has made several contributions relatively recently that bear on some of the issues we have been discussing."

"Yes, please go on," the lawyer requested.

"First of all," Dr. Yardley stated, "Eschenmoser constructed a molecule, known as pyranosyl RNA. This compound contains a modified form of naturally occurring ribose.

"The ribose that occurs in normal RNA contains a five-member ring, consisting of 4 carbon atoms and one oxygen atom. The ribose molecule that forms part of Eschenmoser's pyranosyl RNA compound has been constructed to allow an extra carbon atom in the ring.

"Like normal RNA, complementary strands of pyranosyl RNA are capable of joining together by means of Watson-Crick hydrogen bonding. Furthermore, the use of pyranosyl RNA, with its modified form of ribose, prevents fewer unwanted variations of nucleoside structure from among the multiplicity of available possibilities than does normal RNA.

"In addition, double-strands of pyranosyl RNA do not twist around one another, as is the case with the normal forms of double-stranded RNA. This quality could be extremely important if enzymes were not available, unlike the situation currently, to unwind these strands so that replication could take place."

"Dr. Yardley, as far as you know, does pyranosyl-RNA exist outside the laboratory?" the defense counsel asked.

"No," the professor admitted.

"Would I be fair in saying, Professor," Mr. Tappin queried, "that although one might agree the pyranosyl RNA molecule that has been created in the laboratory is very interesting and suggestive of possibilities, nevertheless, this molecule really is of little practical import to origin-of-life issues if it, or something similar to it, did not exist in the Archean era?"

"Yes, this would be a fair way of saying things," agreed the professor.

"Moreover," added Mr. Tappin, "even if one were to suppose such a molecule as pyranosyl RNA existed in prebiotic times, one would have to explain why, and how, a molecule ... namely, normal RNA ... which, from a number of different perspectives, did not have anywhere near the selective value of pyranosyl RNA, would have come to replace the latter molecule. Would you say these are fair issues to ask?"

"I would assume so," the professor offered.

"Can either of these problems be resolved at the present time," inquired Mr. Tappin.

"Not satisfactorily," responded the professor.

"You stated earlier Dr. Yardley that this fellow Eschenmoser had made several contributions that bear on the issue being discussed. What is the other one?"

"Around 1994," said the professor, "Eschenmoser discovered a way of limiting the kinds of sugars that are synthesized during the formose reaction. Without getting into the technical details of the experiment, essentially, he replaced one of the normal intermediates of the formose reaction with a similar phosphorylated molecule, and, then, he permitted the subsequent steps of the reaction to proceed as normal."

"Excuse me," Dr. Yardley, "am I right in believing that a phosphorylated molecule is a compound to which a phosphate group has been added and that, under certain circumstances, might be capable of storing energy if particular kinds of pyrophosphate bonds are present?"

"Essentially, yes," the professor said.

"Under certain conditions, when this kind of substitution was made, the primary end product of the formose reaction was a phosphorylated derivative of ribose. This substitution process, therefore, represents a possible way of getting around the selectivity problem that arises as a result of the multiplicity of competing sugar forms that exists when one permits the formose reaction to proceed as usual."

Checking the papers in his hand before speaking, Mr. Tappin said: "In the experiment just described, Professor, wouldn't the phosphate group on the synthesized ribose derivative have to be rearranged upon completion of Eschenmoser's altered pathway for the formose reaction in order to be the same as the phosphorylated ribose that is found in normal nucleotides?"

"Yes, that's true," the professor acknowledged.

"In addition," the defense counsel observed, "doesn't the Eschenmoser experiment leave one with a slight problem of needing to explain how one is going to bring about this substitution process under prebiotic conditions when, presumably, there is no Archean era counterpart to Albert Eschenmoser, or his lab assistants, who would be available to make the substitution? Moreover, doesn't all of this assume that the closely related phosphorylated molecule that is to be substituted for the normal intermediate of the ribose-forming reaction is going to be available to be inserted into the formose reaction at just the right moment?"

"I guess so," replied the professor.

"Would you agree, Dr. Yardley," queried the lawyer, "that although there has been some success in synthesizing adenosine and guanosine nucleosides when purified mixtures of ribose and purine bases have been heated in the presence of certain inorganic salts, these same successes are not observed with pyrimidine nucleosides, such as uracil and cytosine, under any conditions that could be considered to be plausible in the Archean era?"

"Yes, that is correct," the professor confirmed.

"Apparently, then," summarized Mr. Tappin, "at the present time there is no known, plausible pathways under prebiotic conditions for synthesizing more than half of the five nucleosides that are fundamental to the storage of genetic information in both DNA and RNA. Is this more or less the state of things in evolutionary theory, Professor?"

"More or less," Dr. Yardley stated.

"To further confuse matters," added the lawyer, "even in the case of the synthesis of the nucleic purine bases, adenine and guanine, one is likely to find other kinds of bases such as hypoxanthine, diaminopurine and a variety of related molecules accompanying the synthesis of the specific purine bases that are important to the nucleic acids that occur in living organisms. So, wouldn't you agree, Dr. Yardley, that, here too, the Archean era, through natural chemical processes, is likely to have generated a variety of cross-linked polymers that somehow would have to be selected against in order to work toward the kind of life form which resembles that with which we are familiar today?"

"Yes, I would agree with this," Dr. Yardley said.

"Would you agree Professor," asked Mr. Tappin, "that all of the problems that have been discussed in relation to the formation of nucleosides would carry over into the formation of nucleotides during which a phosphate component is added to the nucleoside combination of ribose and one of the five nucleic bases? In other words, wouldn't there be a substantial array of abnormal nucleotides consisting of various pentoses other than ribose, as well as forms of ribose other than the right-handed optical isomer of beta-ribofuranose, and, if this is the case, wouldn't these interfere with both catalytic processes as well as RNA replication?"

"Yes, one would have to assume this very well could have been the case," affirmed the professor.

"Dr. Yardley, beside the abnormal nucleotides that would form as a result of the presence of different pentoses, ribose forms and optical isomers, wouldn't there also be an assortment of abnormal phosphate bonds that could arise? In other words, isn't it true that beyond the normal, 5-prime- phosphate bond that occurs during one of the stages leading to the formation of the sorts of nucleic acid found in living organisms, one also might obtain problematic bonding arrangements such as: 2-prime-phosphate bonds; or, 3-prime-phosphate bonds; or, 2-prime-3 prime-cyclic phosphates; or, 2-prime-5 prime-biphosphate; or, 3-prime-5-prime-biphosphates?"

"This is true," affirmed the professor.

"Would you also agree, Dr. Yardley," added Mr. Tappin, "that, in the light of current knowledge, the Archean era is much more likely to have consisted of such a mixture of phosphate bonds, pentoses, different forms of ribose, as well as a racemic aggregation of optical isomers, rather than having consisted of the purified solutions with which laboratory experiments are run?"

"Yes," said the professor.

"In addition, Dr. Yardley, would you agree that despite all the problems that exist in relation to the formation of ribonucleic acids, nevertheless, RNA is more easily synthesized than is deoxyribonucleic acid? In fact, can we not say that one of the considerations that led to the rise of the RNA-world hypothesis was rooted in the way RNA is much more easily synthesized than is DNA?"

"The answer to both of your questions is 'yes'," responded the professor.

"In your opinion, Dr. Yardley," queried the defense counsel, "even if much of the RNA-world hypothesis turned out to be true, wouldn't evolutionary theorists still be faced with the problem of proposing a plausible prebiotic mechanism for the synthesis of DNA?"

"I believe this would be the case, yes," the professor admitted.

"On the other hand," Mr. Tappin indicated, "although RNA is more easily synthesized than DNA, DNA is much less susceptible to hydrolysis, or breakdown in an aqueous environment, than is the case with RNA. If my information is correct, isn't it true, Professor, that at room temperature RNA breaks down at a rate that is roughly 100 times faster than does DNA, and, within certain limits, this differential rate of breakdown climbs somewhat with increases in temperature above room temperature?"

"This is basically right," stated the professor, "except that depending on the temperatures you are talking about, both DNA and RNA tend to decompose more readily at elevated temperatures."

"Dr. Yardley, assuming my understanding of things is right, if one starts with a single polymer or chain of RNA in solution, a complementary strand easily can be generated by adding free, unpolymerized nucleotides to the solution, since, subsequently, these free nucleotides will line up opposite their pairing partner on the original RNA strand ... that is, uracil with adenine and cytosine with guanine. Moreover, the original strand and its complement will form, in the absence of enzymes, a double helical structure by means of the spontaneous hydrogen bonding of these Watson-Crick pairings. Is all of this correct?"

"Yes," the professor replied.

"Yet," the defense counsel stipulated, "the foregoing scenario assumes, does it not, Professor, that all of the free nucleotides that are being added to form the complementary strand must exhibit the same optical properties or handedness as the original strand of RNA?"

"That's correct," Dr. Yardley affirmed.

"In other words," indicated the lawyer, "if one places both left- handed and right-handed optical isomers of various free nucleotides into the solution, then, the presence of both left- and right-handed isomeric forms of the nucleotides will inhibit the formation of a complementary strand capable of bonding with the original strand through Watson-Crick pairings. Isn't this so, Dr. Yardley?"

"Yes, it is," acknowledged the professor.

"Furthermore," Mr. Tappin continued, "according to the information that is available to me, despite years of experimental efforts by hundreds, if not thousands, of scientists and researchers, no one has been able to find a way to replicate or copy a complementary strand of nucleic acids without the assistance of enzymes. Consequently, would you agree Dr. Yardley, that although scientists can generate, in the absence of proteins, a complementary strand for an original strand of RNA, these same scientists cannot copy the complementary strand without the right kinds of enzyme being present?"

"Although, in general, much of what you have said is true," the professor indicated, "I wouldn't agree with your statement without adding at least one qualifying remark. More specifically, two researchers, by the name of McHale and Usher, have demonstrated that when strands of RNA oligonucleotides, consisting of 10 polymerized units or less, are dried and heated in temperatures that approximate sunlight, these RNA oligonucleotides will line up along a complementary template and form polymers or bonded chains similar to the process of replication that occurs in living cells."

"Correct me if I'm wrong, Dr. Yardley, but I believe," suggested the lawyer, "there are a number of differences between the experiment you are describing and the conditions one is likely to be working with in an Archean era environment. First of all, wouldn't you agree, Professor, the experiment to which you are alluding is presupposing what has not, yet, been able to be satisfactorily demonstrated by evolutionary science -- namely, that normal RNA nucleotides would have been synthesized and selected out in pure, concentrated forms from amongst the motley array of possibilities involving: pentose sugars, different forms of ribose, optical isomers, alternative phosphate bonding possibilities, lack of pyrimidine bases, as well as a variety of odd purine bases in addition to adenine and guanine?"

"Yes, that is correct," the professor responded.

"Isn't it also the case, Dr. Yardley," queried the defense counsel, "that in living cells there is an unwinding protein that is able to help separate the individual strands of the double-helix form of nucleic acids that is being held together by Watson-Crick hydrogen bond pairings. In fact, in your previous discussion of Eschenmoser's laboratory creation, pyranosyl RNA, wasn't one of the attractive features of this molecule the fact it offered a possible way around needing a protein to unwind the double-helix structure of nucleic acids?"

"That's right," said the professor.

"Consequently, isn't the McHale-Usher experiment presupposing," Mr. Tappin asserted, "that there was a means, under Archean era conditions, to unwind the strands that spontaneously tend to form double-helix structures through Watson-Crick pairings in order for there to be a complementary, single-stranded template with which to work?"

"This would seem to be the case," the professor agreed.

"To the best of your knowledge, Dr. Yardley, has any ribozyme ... that is, an RNA polymer with catalytic activity -- been discovered that has the required unwinding capacity that appears to be presupposed by the McHale-Usher experiment?"

"Not as far as I know," answered the professor.

"Furthermore," the lawyer added, "given that the experiment was successful with short polymers of 10 units or less, one is left wondering why the same kind of experiment has not been successful in the replication of much, much longer polymers of nucleic acid as would be required in fully functioning, living cells. In fact, Professor, isn't it the case that part of the lack of experimental success with respect to being able to polymerize long sequences of RNA molecules is due to the instability of the RNA molecule? In other words, isn't it true that the rate of RNA polymerization must take place fast enough to compete with the rate of random, hydrolytic decomposition of the same RNA molecules, and this is difficult to achieve in the absence of protein enzymes that have the capacity to increase reaction rates by magnitudes of between one million and one billion times?"

"Yes, I guess so," responded the professor, "but, if nothing else, I believe the McHale-Usher experiment is very suggestive and carries a lot of implications for the origin-of-life issue."

"Finally, Dr. Yardley, wouldn't you agree," Mr. Tappin inquired, "that the experiment in question is assuming the following. Even if one, or more, normal RNA oligonucleotides somehow found their way into existence under Archean era conditions, nevertheless, the researchers do not seem to be allowing for the possibility of the degradation or decomposition of these molecules through hydrolysis, ultraviolet photolysis or pyrolysis?"

"Quite frankly," replied the professor, "I' m not sure I would agree the researchers should have to take any of these factors into consideration. The experiment was intended to show a possibility rather than be a definitive way of resolving all conceivable problems facing evolutionary theory."

"Fair enough," responded the defense counsel, "but would you agree, in turn, that even if McHale and Usher do not have to take any of these various, nevertheless, if evolutionary theory is to provide a plausible account for the origin-of-life through natural processes, then, this theory must be able to resolve the problems that are being raised in relation to the McHale-Usher experiment. After all, just as there are positive implications that follow from the McHale -Usher experiment, are there not also a number of negative or problematic implications that are inherent in that same experiment?"

"I guess I can live with this way of stating things," offered the professor.

"During direct examination testimony, Dr. Yardley, you spoke about a number of different ribozymes or sequences of RNA with catalytic properties. If I remember correctly, these properties involved such activities as the cutting and splicing of specific RNA sequences, as well as assuming some limited characteristics of a polymerase by helping to bring about the formation of the bonds that link together certain kinds of polymer chains. Is this right?" the lawyer asked.

"Yes," affirmed the professor.

Mr. Tappin briefly looked through the material he had been holding in his hands while conducting the cross-examination. After five or ten seconds of searching, he pulled out a sheet of paper and placed it on top of the material in his hands.

Eventually, he said: "Dr. Yardley, in doing research concerning some of the experiments dealing with ribozymes, I came across something about which I'm curious. Perhaps, you can help me out.

"At one stage during the particular study that I have in mind," explained the lawyer, "the researchers were interested in determining whether the catalytic specificity exhibited by a naturally occurring ribozyme could be overcome or altered. More precisely, these researchers wanted to see if the ribozyme could be induced to interact equally effectively with a variety of base sequence combinations rather than just the limited nucleic sequences for which the ribozyme, under normal circumstances, seemed to show an inherent, interactive preference.

"In order to overcome the inherent sequence specificity of the ribozyme, the researchers began exploring the possible effects that a variety of polyamines might have on the ribozyme. Although, undoubtedly, Professor, you know what a polyamine is, for the benefit of the jurors, a polyamine, as the name suggests, is a compound that contains two or more amino groups.

"Now," the defense counsel continued, "the simplest of polyamines, such as putrescine [$NH_2(CH_2)_4NH_2$] and spermine [$NH_2(CH_2)_3NH(CH_2)_4$-$NH(CH_2)_3NH_2$] are far more complex than compounds such as hydrogen cyanide (HCN), methane (CH_4), formaldehyde (CH_2O), or ammonia (NH_3). Yet, there is considerable discussion concerning the extent of the availability of even these latter, simple hydrocarbons during Archean era times.

"There were ten polyamines that were tested during the experiment. Only one of these polyamines, spermadine, which is of moderate complexity relative to other polyamines, was found to be capable of inducing the ribozyme to overcome its inherent base sequence specificity.

"Once again, Professor, as was true in relation to the original origin-of-life experiment of Miller, or any of Fox's proteinoid experiments, or Eschenmoser's pyranosyl RNA molecule, and numerous other experiments that supposedly simulate the conditions of the prebiotic Archean era, I question the value of such experiments as far as their implications for origin-of-life issues are concerned. How much spermadine, Dr. Yardley, was there in the Archean era world?"

"The short answer to your question," replied the professor" is that I don't know. Although polyamines might be more complex than the simpler compounds from which various origin-of-life scenarios usually begin, the quality of complexity does not, in and of itself, automatically preclude the possibility that polyamines could not have been synthesized under prebiotic conditions.

"As I indicated previously," pointed out the professor, "just because an experiment is performed that does not necessarily faithfully simulate certain aspects of the conditions of the Archean era, this does not mean such an experiment cannot have implications for what might have gone on during prebiotic times. For example, even if one were to assume that spermadine didn't exist during the Archean era, the fact that, under certain conditions, ribozymes can be induced to broaden their catalytic activity, raises the possibility there might have been other agents that did exist during the Archean era and that might have had an effect on ribozymes similar to the action of spermadine.

"If we didn't know about what spermadine helps make possible, we might not have a reason to go looking any further to determine whether there might have been a more plausible prebiotic method for bringing about the same kind of result that spermadine does. In all likelihood, the experiment to which you refer was not, in any technical sense, intended to serve as a simulation experiment, but, nevertheless, this experiment provides evidence that helps shape theory and future experiments as well as strengthens the overall evolutionary model."

"Would you say, Dr. Yardley that the spermadine experiment constitutes evidence in support of evolutionary theory?" Mr. Tappin inquired.

"If you are asking me," the professor replied, "whether this experiment constitutes a sort of 'smoking gun' that brings us to the brink of completing an unbroken chain of evidence that overwhelmingly and undeniably demonstrates the truth of an evolutionary explanation for the origin-of-life, then, my answer is the spermadine experiment does not provide the kind of evidence in support of evolutionary theory that you are seeking. If, on the other hand, you are asking me whether the spermadine experiment provides information that helps to shape, color, modulate, and orient evolutionary theory, then my answer is that this experiment does constitute evidence in support of evolutionary theory."

"Actually, Dr. Yardley," Mr. Tappin responded, "I'm asking neither kind of question. The question that I'm posing is more like the following: given that legitimate questions can be raised about the availability of polyamines such as spermadine in the Archean era, does the fact a ribozyme can be experimentally induced to overcome its inherent sequence specificity under artificial, and prebiotically unrealistic, conditions, really bring us any closer to answering the question of how life came into being, especially in view of the very strong possibility that ribozymes might not have been capable of being synthesized in the prebiotic world?

"In other words, Professor, many evolutionary researchers seem to be saying: if such and such a set of conditions holds, then, such and such an outcome is possible, and if we assume that these condition s did hold during the Archean era, then, this constitutes evidence in support of evolutionary theory. Yet, the question that really needs to be asked and answered is this: do we have any plausible means of demonstrating the likelihood that such a set of conditions existed and that such an outcome did, in fact, take place during the Archean era?"

"All of evolutionary theory," Dr. Yardley asserted, "is about establishing and demonstrating how some conditions, events, processes and outcomes might have been more likely than other conditions, events, processes and outcomes."

"That might well be true, Professor, but there seems to be a heavy fog warning that is being posted with respect to conceptual travel in the areas of 'demonstration' and 'likelihood'," the defense counsel replied. "For instance, you previously said the spermadine experiment can be considered to constitute evidence in support of evolutionary theory because, irrespective of whether it is right or wrong, the findings of the experiment can be used to help shape and modulate that theory, and, yet, at the same time, the spermadine experiment might have nothing to do with the Archean era, and, therefore, by implication, the spermadine experiment might have nothing to do with one of the most important questions facing evolutionary theory ... namely, how did life come into being.

"In effect, I'm having a little trouble, Dr. Yardley, understanding how you propose to reconcile these seemingly antagonistic elements. If, and the viability of this 'if' needs to be examined ... if one can raise questions that cast serious doubt on the degree of relevance of the spermadine experiment with respect to helping us resolve the origin -of- life issue, then, how does it serve as evidence for evolutionary theory?"

"Science," suggested the professor, "is about empirically and conceptually exploring possibilities concerning the physical/material world in a methodical, rigorous fashion. Within certain limits, whatever an experiment permits us to eliminate in the way of possibility, we eliminate. Similarly, within certain limits, whatever an experiment permits us to retain in the way of possibility, we retain.

"Over time, the relationship between what has been eliminated and what is retained takes on a structural form. We describe this relationship through the concrete vocabulary of hypothesis, conjecture, experiment, methodology, data, evidence, analysis, principles, laws, theory, and model.

"Unfortunately, at any given time, there is often a certain amount of ambiguity that surrounds the issue of what justifiably can be eliminated or retained as a function of the empirical data and experimental results that might be in our possession. The spermadine experiment gives expression to a certain amount of this sort of ambiguity.

"On the one hand, as you rightly point out" affirmed the professor, "we don't know whether spermadine, or ribozymes for that matter, existed during the Archean era, although there is evidence that can be offered both for, and against, such possibilities. Even if we eliminate the ontological possibilities of spermadine and ribozymes from the picture, we still can retain the idea that something like them might have existed and which, if they did, would help resolve certain kinds of problem, so, we proceed to try to determine whether we should eliminate or retain such conceptual possibilities on the basis of forthcoming empirical data and conceptual reflection.

"On the other hand, if spermadine and ribozymes did exist during the Archean era -- a possibility concerning that, once again, evidence can be offered both for and against ... then, the spermadine experiment is revealing a very interesting possibility that ought to be retained and explored further. Now, although the available evidence does suggest there are a variety of factors that help mitigate against continuing to retain either spermadine or ribozymes as viable, plausible pieces of the origin - of-life puzzle, nonetheless, we have not yet reached a point where these possibilities can justifiably be eliminated from the picture.

"Quite frequently, there is a constant dialectic and tug-of-war going on between how we feel about what, both conceptually and empirically, should be eliminated and what should be retained at any given time. Consequently, despite the fact something might have a theoretical status, vis-à-vis elimination and retention, which is ambiguous, nonetheless, such an ambiguous element still can come to have a shaping influence on one's theories, models, conjectures and hypotheses, even while there are other factors that serve as contraindications to this shaping influence."

"What happens," hypothesized the lawyer, "if your feelings about the proper relationship between what is to be eliminated and what is to be retained are at odds with my feelings about the proper relationship between what is to be eliminated and what is to be eliminated?"

"Then," the professor said with a shrug of his shoulders, "we have a difference of opinion."

"Is there any way to resolve such a difference of opinion," the defense counsel asked.

"Yes and no," answered the professor. "One can try to do more science until the balance of evidence seems to point more in the direction of one kind of relationship of elimination/retention rather than some other such relationship. However, this often is easier said than done, and, moreover, there frequently are other ideas about the proper relationship between what should be eliminated and retained that arise in the meantime and complicate any straightforward resolution of the original difference of opinion.

"Progress does occur in the sense that despite a variety of differences of opinion about numerous issues concerning what should be retained and what should be eliminated, a broad consensus develops about some of the things, both empirical and conceptual, that should be eliminated and some of the things that should be retained. Even here, however, one finds some people who are resistant to either eliminating possibilities or retaining possibilities despite the presence of a general consensus among many researchers on such matters."

"Does the existence of a consensus," queried the lawyer, "necessarily mean this decision on what, in broad terms, should be eliminated or retained is, in some sense, a correct one?"

"Not at all," Dr. Yardley stated. "Yet, one could say that where such consensus exists, there usually is considerable justification that can be offered ... through empirical observations, experimental results and conceptual analysis ... in support of such decisions, and, therefore, anyone who wishes to oppose these kinds of decision will be swimming against the tide of an informed consensus of opinion.

"Of course, historically, conceptual revolutions often have come in the form of one or more people who believed the wrong consensus decisions had been made about the possibilities that are being eliminated, retained or even entertained. Apparently, your client, Mr. Corrigan, is an individual who feels consensus opinion concerning evolutionary theory is wrongheaded, but whether his opposition will result in a revolution or merely fall by the wayside as a very minor historical oddity will be decided, to some extent, by what the present jury and other similar forums of public opinion decide."

"I've noticed," Mr. Tappin observed, "there doesn't seem to be a lot of talk about the notion of truth in your characterization of science. Given that many people normally link issues of scientific evidence and demonstration with the idea of having, to some extent, proven that something is true, I'm wondering if you might elaborate a little on this aspect of science."

"Naturally," Dr. Yardley replied, "researchers hope that, in some way, elements of reality are faithfully captured in what is retained by the scientific community. Similarly, researchers hope everything that we eliminate is being thrown out because it lacks this quality of faithfulness or reflectivity when compared with experience, experiment, analysis and so on.

"In fact, generally speaking, there are only two kinds of mistake that can be made in science. On the one hand, we can retain something that, in reality, turns out to be incorrect, erroneous, false, and, therefore, in some sense, distortive with respect to our experience concerning what is. On the other hand, we can eliminate something that, in reality, turns out to correct, accurate, true, and, therefore, is, in some sense, reflective of our experience of what is.

"The problem in all of this is that, quite frequently, there are distortive elements mixed in with the reflective features that are retained, just as there often are reflective elements mixed in with the distortive features that are eliminated. This adds to the ambiguity of the situation to which I alluded earlier, and this also helps to explain why researchers are not inclined to rush to judgment about what should be retained or eliminated, and also why some individuals are reluctant to eliminate certain possibilities despite a contrary judgment by the consensus of opinion of the scientific community.

"Oddly enough, at least from the perspective of some people, scientists are more inclined to want to talk about the beauty of a theory rather than its truth. Etched deep in the psyche of many a scientist is the belief that whatever truth or reality might ultimately turn out to be, it will be beautiful as well.

"Because the truth is not always easy to come by or discover, scientists often use the beauty of a theory as a possible index or sign of the presence of truth within the theory. Like so many bag-people, researchers furiously rifle through the garbage cans of empirical data in search of the nuggets of truth that are to be retained while we wait for the dump trucks of history to remove the remaining refuse, and, often times, the only thing that sustains our search is the beauty of the receptacles through which we are foraging and the belief that such beauty is, at least in part, derived from the sweet smell and colors of truth contained somewhere in the garbage cans through which we are searching."

"What is meant by the notion of the beauty of a theory," the defense counsel inquired.

"The beauty of a theory is not always easy to pin down. A lot of the time, researchers recognize such beauty when they encounter it, but they would be hard pressed, if asked, to delineate the nature of such beauty prior to, and sometimes even after, the actual encounter experience.

"There are, however, some classic indices usually associated with the beauty of a theory. For instance, a beautiful theory often tends to be able to lend a directed and consistent sense of meaning and organizational orientation to disparate sets of data, observations, ideas, experiments, and findings.

"Normally speaking, the data of life look like a scatter diagram with the temporal, spatial and qualitative co-ordinates of experience appearing as just so many unconnected and unrelated points. Then, someone comes along with a theory that shows a way of connecting many of the plotted points of experience in a very consistent, meaningful and organized manner, sort of like when one comes up with a regression line to give linear expression to the various tendencies contained within the scatter diagram at which one has been staring and trying to make sense out of its many data points.

"When one sees conceptual order emerge out of seeming chaos and disorder, the experience is a very aesthetic one. The beauty being given expression through this aesthetic dimension is very compelling and alluring.

"Another qualitative index of a beautiful theory revolves around the notion of simplicity. The capacity of a theory to take a few fundamental ideas and weave them together into complex patterns that can encompass an ever-expanding horizon of experiences, possibilities, and so on, has the aura of beauty about it.

"No matter how complicated things become, one always can return to the few simple ideas out of which the theoretical tapestry has been woven and, thereby, develop a deep aesthetic appreciation for how the whole pattern has arisen as a function of those underlying ideas. Under such circumstances, one's understanding might be fuzzy with respect to the details and minutiae of theoretical complexity, but grasping the simple elements and forces that bind, and animate, the complexity, allows one to be able to orient oneself in the midst of uncertainty.

"This dimension of simplicity has a quality of beauty about it. When researchers encounter this property, we tend to be very attracted by it.

"A third index of a theory's beauty revolves around the heuristic value and power of such a theory. This quality is intimately connected to the two previous facets of theoretical beauty, namely its dimensions of simplicity and organizational capacity.

"When one combines organizational strength with simplicity, this tends to lead to a conceptual dialectic and dynamic that becomes very fruitful with respect to the possibilities, ideas, experiments, hypotheses and explorations that are set in motion by this kind of dialectic and dynamic. The more fruitful a theory is in these respects, the more powerful, stimulating, productive, and valuable the theory becomes.

"This heuristic component of a theory -- that is, its conceptual and experimental fruitfulness, and, therefore, its power ... is, obviously, very desirable. When researchers encounter it, we tend to find it to be a thing of beauty.

"A fourth index of beauty in scientific thinking revolves around the notion of symmetry. This property deals with the capacity of a theory to allow different parameters and variables within that system to undergo operational transformations without the essential aspects of the theory being altered, so that observers in various frameworks will agree these essential features remain the same across the transformations, and, therefore, those features are considered to have been conserved.

"Finally," the professor concluded, "there is an aura of integrity and nobility about a theory that possesses beauty. A beautiful theory tends to stand against the onslaught of confusion, error, darkness, ignorance, and corruption that surround us ... repelling, in an eloquent and elegant fashion, the potential forces of conceptual and social dissolution.

"All in all, the aesthetics of a beautiful theory allow researchers to develop a feeling for some of the realities with which they are attempting to deal. By following this aesthetic pull, researchers are quite frequently led to closer approximations of, or better reflections of, the truths that often are aligned closely to the presence of beauty in a theory.

"I suppose, in many ways, researchers believe it is not possible for a theory to exhibit the various dimensions of beauty, such as organizational meaning, simplicity, heuristic value, symmetry and integrity, without the truth being involved in some fashion. Consequently, seen from this perspective, science really becomes a rigorous, methodical exploration for the elements of truth or reality that researchers believe are being reflected in, and, consequently, that are responsible for, a given theory's beauty."

"Dr. Yardley, couldn't one argue," Mr. Tappin postulated, "that throughout history, including the history of science, there have been a succession of aesthetic theories of truth, if you will, which have been quite captivating and alluring during their time, but, with the passage of time, the beauty of these theories has faded?"

"Yes, this frequently has been the case," acknowledged the professor.

"Moreover," the defense counsel continued, "don't we all, whether or not we are scientists, constantly have to grapple with the possibility that what we find beautiful might, in reality, be a counterfeit, or an illusion, or purely a subjective projection being imposed onto the character of experience or reality?"

"Yes," the professor said.

"Furthermore, Dr. Yardley, would you agree," the lawyer asked, "that, perhaps, on occasion, the reason why we find a theory beautiful is because it serves our personal interests, needs and aspirations, rather than because the theory's beauty is an index for, or sign of, the presence of truth."

"Again, I would agree, in principle, with what you are saying," affirmed the professor.

"In addition," Mr. Tappin pressed, "isn't it possible that what we take to be the reflective beauty of truth and reality is but the reflection of a scientific, political, religious, cultural and/or philosophical conception of beauty and truth into which we have been initiated or indoctrinated by the formal and informal aspects of the educational processes to which we have been exposed during our lives?"

"Of course, this is a possibility," remarked the professor.

"Lastly, Dr. Yardley, don't myths have many of the same kinds of properties that you have outlined with respect to the idea of beauty?

In other words, don't myths have the capacity to offer organized systems of: directed meaning, simplicity, heuristic value, symmetry, and a certain kind of integrity and nobility of purpose?"

"Yes, I suppose so," the professor responded, "but I believe the qualities of beauty in science are a lot more sophisticated, methodologically sound, and analytically rigorous than anything that might be generated through myths."

"Maybe you feel this way, Dr. Yardley, because you are firmly caught up in the myths of science. Isn't this possible?"

"Perhaps," stated the professor.

Reviewing the material in his hands, Mr. Tappin asserted: "In earlier testimony, we have established that, so far as is known, there is no ribozyme capable of unwinding double helical structures that have assumed a stable state through Watson-Crick pairing. In similar fashion, Professor, is there any naturally occurring ribozyme that has proven to be capable of serving as the RNA-world's counterpart to the exonuclease proteins that are able to eliminate errors during the replication of nucleic acid polymers?"

"Not so far," Dr. Yardley indicated.

"What happens if there is no means of maintaining replicational fidelity from one generation to the next?" Mr. Tappin asked.

"Within limits," Dr. Yardley pointed out, "a system can tolerate a certain amount of replicational infidelity. A lot depends on where such errors occur since some pathways and functions are a lot more crucial than are others.

"In addition, under some circumstances, errors in replication actually serve a positive function. Such errors become the mutations through which new evolutionary possibilities might be introduced into the system.

"However, when the replicational fidelity of a genetic system falls below a certain level, then, vital information is lost, not only with respect to the individual, but also in relation to the species population as well. Generally speaking, any kind of replicational process that falls much below, say, a 96-99 percent fidelity rate per nucleic acid residue is very likely, sooner or later, to run into problems that will challenge the continued existence of the kinds of pathways, reactions, structures, activities and functions that are being underwritten by such a replicational process."

"If the RNA-world hypothesis is to be taken seriously," postulated the defense counsel, "wouldn't it have to be able to propose some plausible way to ensure that the fidelity of replication from one RNA generation to the next could be maintained? In fact, wouldn't such a capacity be of the utmost importance given the vast range of abnormal nucleotides and nucleosides that are likely to be roaming about in an Archean era environment?"

"Yes," agreed the professor, "an exonuclease -like capability would be very important to an RNA-world, just as such a capacity is crucial to the DNA-world in which we live."

"I'm sorry, Professor, could you briefly explain what an exonuclease is," Mr. Tappin requested.

"Perhaps, the easiest way to describe the function of this kind of molecule" responded the professor, "is to say they are able to identify and eliminate the vast majority of errors that might arise during, say, the process of replication."

"Thank you," the lawyer acknowledged, and, then, he proceeded to ask: "Can one assume, Dr. Yardley, that a plausible RNA-world hypothesis would require substantially fewer kinds of functions ... such as, but not limited to, the just mentioned exonuclease ... than the DNA- world requires in the way of structural and enzymatic proteins?"

"No, I wouldn't think so," the professor replied.

"Yet," challenged the defense counsel, "only a very few, limited ribozymes have been discovered so far. How do these few discoveries lend much plausibility to a RNA-world hypothesis?"

"First of all," Dr. Yardley responded, "these discoveries are important because of their implications. The fact there might be few ribozymes in existence today does not preclude these molecules from having been a dominant force at some early stage of evolutionary history.

"Secondly, and related to the first point, the ribozymes we have been finding might merely be the left-over remnants of the order of things that once was, just as our appendix might be an evolutionary remnant of an organ or process that once had a function at some point in our evolutionary past. These sorts of evolutionary relic are found throughout the animal and plant worlds.

"Thirdly, the discovery of ribozymes opened up a lot of conceptual possibilities that helped set the stage for a variety of exploratory probes, both experimental and theoretical in character. A lot of important work has come out of the RNA-world hypothesis that has helped to expand the horizons of the evolutionary model in a number of ways.

"Admittedly, there are quite a few outstanding problems facing the RNA-world hypothesis. However, even if this hypothesis is eventually rejected or abandoned, science and evolutionary theory will have benefited by going through the rigorous processes of questioning, experimenting, analyzing, and reflecting that have been necessary in order to properly consider the possible tenability or value of such a hypothesis."

"Gentlemen," interjected Judge Arnsberger, "I feel the time has come to put the discussion to bed for the night. We'll pick things up again tomorrow morning at 10:00 a.m.

"I trust the jurors will continue to behave themselves with respect to the restrictions that have been placed on their discussing the case with anyone. Court is adjourned."

Transposable Conceptual Elements

Mr. Tappin studied the papers in his hands for about five or ten seconds. When he had finished, he asked: "In the Cech and Zang study involving a particular kind of ribosomal activity, one comes across references to something known as 'L -19 IVS RNA'. What is this?"

"This is the working name," Dr. Yardley explained, "for a large molecule of ribosomal RNA. The L-19 portion of the designation refers to the 19 nucleotides that have been removed from an original sequence of 395 nucleotides by the catalytic self-splicing action of this molecule.

"Because the original sequence catalytically operates on, or intervenes with respect to, itself, it is referred to as an intervening sequence. This is the IVS component of the working name."

"What function is served when the 395-nucleotide polymer cuts off 19 nucleotides from itself?" the lawyer inquired.

"Apparently," replied the professor, "this provides a more accessible binding site on the L-19 IVS RNA molecule to which several other oligonucleotides, or short sequences of nucleic acid, can be brought together to form a bond through what is known as a transesterification reaction. In effect, the L-19 IVS RNA enhances the rate of hydrolysis that is characteristic of this sort of reaction by a factor of 10^{10} ... or 10 billion times.

"This kind of transesterification reaction has never been observed to occur between two free oligonucleotides. Consequently, the presence of a protein enzyme or, as in the present case, an RNA ribozyme is of paramount importance if such reactions are going to occur."

"How large," asked the lawyer "is the binding site that is made available by the cleaving of the 19 nucleotides from the original 395 nucleotide IVS RNA molecule?"

"We believe it to be about 7 nucleotides, or so, in length," the professor answered.

"If the binding site is only 7 nucleotides in length," the defense counsel reasoned, "why is there a need for the other 388 nucleotides? Why doesn't the original IVS RNA molecule simply cleave off all but the 7 nucleotides that constitute the binding site?"

"First of all," pointed out the professor, "the 395-nucleotide sequence supervises the initial, precise process that eliminates the 19 nucleotides that render the binding site more accessible to the nucleotides that are to be chemically bonded together. Secondly, the remaining L-19 IVS nucleotide sequence also supervises, so to speak, the bringing together of nucleotides and, in doing so, is required to recognize three or more nucleotides in order to establish a reaction site.

"Consequently, the L-19 IVS RNA molecule has more base-sequence specificity for single-stranded RNA than many, if not most, protein enzymes that are involved in similar kinds of reactions under other cellular circumstances. In fact, this specificity might even rival the specificity of various DNA restriction endonuclease protein enzymes that key in on, and cleave, very specific bonds such as those occurring during the unwinding process of the double-helix structure that is preparing for replication.

"Various kinds of base-deletions studies have been done in relation to IVS RNA to determine just how much of the original 395 nucleotides are necessary for efficient cleavage-ligation activity. On the basis of these kinds of study, at least 300 nucleotides appear to be minimally required in order for efficient catalytic activity to be manifested."

"Does this mean" the defense counsel queried, "that all ribozymes would have to be this large in order to be effective catalysts?"

"At this point," the professor indicated, "we are not quite sure.

There are molecules known as group-I introns whose core structure consists of about 100 nucleotides and that exhibit considerable catalytic activity.

"As a result, seemingly, not every ribozyme necessarily has to be as big as, say, the 300 nucleotides that appear to be minimally necessary for effective IVS RNA functioning. There might be a range of possible ribozyme sizes depending on function and so on, but, at the present time, we do not know what the upper and lower limits of this range might be."

"Given the catalytic specificity of these ribozymes," postulated Mr. Tappin, "even if we were to select, say, a group-I intron consisting of 100 nucleotides, wouldn't the odds of generating this kind of specific sequence on a random basis be, at a minimum, 4^{100}, since there are four nucleic bases that could occupy any one of the 100 nucleotide positions in the entire sequence?"

"Yes, this is correct," the professor confirmed.

"Similarly, for the, let us say, 300 nucleotide IVS RNA molecule," the defense counsel added, "the odds of generating such a specific sequence on a purely random basis would be 4^{300}. Is this right, Dr. Yardley?"

"Yes," said the professor.

"Previously, Dr. Yardley, you have suggested the entire Archean era was filled with mini-prebiotic laboratories. Let us suppose we were to give those laboratories about 400 million years to come up with the correct sequence for a ribozyme consisting of 100 nucleotides -- the 400 million years being near the figure you cited in direct examination testimony for the length of time during which life is likely to have originated on Earth.

"Let us further suppose all activity in these mini-prebiotic laboratories stopped except work that was directed toward coming up with the right sequence for one specific ribozyme catalyst consisting of 100 nucleotides. How many experiments, Dr. Yardley, would have to be performed per day, over the course of the allotted 400 million years, in order to exhaust the 4^{100} combinations of nucleotide sequences that are possible?"

The professor was silent for about 15 seconds and, then, said: "Probably, in the vicinity of 3×10^{88} experiments per day."

"I've read somewhere, Professor," stated the lawyer, "I forget where, that the surface of the Earth covers about 196,938,800 million square miles. Assuming this figure to be correct and if we were to assume that every square mile of the Earth was to be dedicated to trying out experimental combinations of 100 nucleotides to come up with the specific sequence of our Group-I ribozyme, how many experiments would have to be performed per square mile in order to exhaust the possible combinations?"

"About 2×10^{92} experiments per square mile," replied the professor.

"Of course," Mr. Tappin indicated," we have been assuming in all of the foregoing that we are dealing with the same kind of nucleotides that occur in living organisms. If we add in the assortment of different pentose sugars, ribose forms, optical isomers, odd nucleic bases, and phosphate bonds that are likely to have been hanging around during Archean era times, then, Professor, won't we have to significantly revise all of the foregoing figures in an upward direction in order to factor in the increased possibilities for combining 100 nucleotides in a specific sequence?"

"Yes, we would," Dr. Yardley responded.

Mr. Tappin held up the papers in his hand. "Professor Yardley, according to the information available to me, an Escherichia coli bacterium contains 4 million base pairs of nucleic acid. Let us assume, arbitrarily, that the first self-sustaining life form had only one-quarter as many base pairs ... that is, 1 million base pairs. Would this be a fair assumption?"

"Nobody really knows," stipulated the professor. "No one knows how few ribozymes or enzymes one needs in order to have a self- sustaining, self-replicating organism or protocell.

"Obviously, one needs more genetic information than is carried by a virus since such entities presuppose the existence of a host's replicating capabilities in order to produce new generations of the virus. However, precisely how much more would be minimally necessary is, at the present time, an open theoretical question."

"Let's assume," the defense counsel proposed, "that the average ribozyme is 100 nucleotides in length. In an RNA-world scenario, how many ribozymes do you feel, Dr. Yardley, would be reasonably necessary to look after the catabolic and anabolic pathways of a minimally functioning protocell capable, I would presume, of, to varying degrees: self-replication, division, growth, membrane transport, ion pumping, energy storage, charge transfer, ribosomal activity and the like?"

"I only would be blindly guessing," the professor stated. "Maybe, somewhere between: 100 and 200 ribozymal genes."

"All right," Mr. Tappin suggested, "let's take the lower boundary figure of 100 ribozymal genes. This means, in effect, the mini-prebiotic laboratories would have to find a collective way of bringing together in one place and at one time, a specific sequence of 10,000 nucleotides, 100 times, which is 1 million base pairs -- the number of base pairs we are assuming to have been in our pre- E. Coli life form -- divided by our arbitrary and average figure of 100 ribozyme genes.

"If we were to assume there were a naturally occurring pathway for synthesizing ribonucleic acids, and if we were to assume there were a plausible means of polymerizing these nucleotides under prebiotic conditions, and if we were to assume there were no cross-bonding of pentoses, odd nucleic bases, phosphates, optical isomers, or different forms of ribose, then there are 4^{10000} possible combinations for a series of sequences adding up to 10,000 polymerized nucleotides. Now, none of the foregoing takes into consideration the fact that even given such a specific sequence of nucleotides, the order in which the ribozymal genes are activated and deactivated, as well as when, or for how long, this process of turning the genes on and off takes place, all of this has to be factored into calculating a baseline probability figure.

"Consequently, the 4^{10000} figure is, very much, a lower boundary figure for calculating the odds of generating such an arrangement of nucleotides if one were to assume chance factors were to be the only determinate in inventing such an effectively, functioning system capable of self-replication. Would you agree with this, Dr. Yardley?"

"These are your figures, Mr. Tappin," indicated the professor, "but, for the sake of argument, I'm willing to live with them."

"Given the foregoing, Dr. Yardley, would you be surprised," asked the lawyer, "if I were to tell you there would have not been enough time, space, energy or organic materials on Earth for the mini-prebiotic laboratories to experimentally search through even an extremely minuscule fraction of the total possible combinations that rise from a protocell organism with a genetic repository of 10,000 nucleotides during the 400,000,000 year, or so, period in which life is thought to have originated according to evolutionary theory?"

"Look," the professor asserted, "sometimes one can get figures and numbers to dance almost any tune one likes. There are many possibilities that are not being taken into consideration by your calculations."

"Such as ..."the defense counsel queried?

"There might have been," the professor proposed, "selective forces and conditions operative in the Archean era that might have placed severe constraints on many of the combinations and, as a result, preempted the need for an extended search. For instance, if some given prebiotic experiment produced results that were compatible with the existing thermodynamic and kinetic conditions of a particular evolutionary niche, then, such a result would tend to be selected over other prebiotic experimental results that either were not compatible with existing conditions, or were not compatible to the same extent and, therefore, were at a selective disadvantage as far as thermodynamic and kinetic forces were concerned.

"If one extrapolates this process across the course of hundreds of millions of years, then, there is a finite, but extremely large, set of intermediate steps, all of which could have been selected by available thermodynamic and kinetic conditions. Looking backward, after billions of these steps have occurred, one might have difficulty in understanding how one has got to where one is.

"One also might become overwhelmed when one considers the vast numbers of prebiotic reactions that were experimented with but that were not compatible with the shifting fortunes of thermodynamically and kinetically favorable conditions. Finally, one might be totally amazed when one performs the calculations and discovers what the odds were against this happening on a purely theoretical basis, but the theory on which such calculations is based has not, and, probably cannot, take into account the way a series of thermodynamic and kinetic conditions have selected for a succession of results that has permitted the improbable to be overcome.

"In addition, you are assuming the entire set of combinatorial possibilities would have to be searched before the correct sequence was found. Conceivably, a functional solution could have been discovered at any juncture of the search, and there is no way of predicting when this juncture will be reached.

"Odds become meaningless to the person who is struck by lightning or who wins a lottery. Similarly, no matter how improbable the theoretical odds are concerning a sequence of 10,000 nucleotides, if the prebiotic version of a jackpot occurs, all we can do is to say that on the basis of theoretical calculations we were not expecting such an ontological event to occur and that one would fully anticipate such a rare event to be extremely unlikely to happen again ... although who knows, lightning sometimes does strike twice."

"Am I to understand, Dr. Yardley," Mr. Tappin wondered, "that if I were to remove all but one bullet from a revolver, spin the chamber, hand it to you, and tell you to point the muzzle toward your brain and pull the trigger five times in succession, you would do so because we can assume the bullet has, relative to the formation of a 10,000 sequence nucleotide, an inordinately good chance of showing up in the last chamber?"

"No, this is not what I'm saying," remarked the professor. "Your counter-example is not the same thing."

"Why isn't it?" inquired the lawyer. "Is it because, my counter- example, unlike your various assumptions about possibilities during the Archean era, is capable of demonstration?

"What evidential, demonstrable, rigorous reasons do I have," Mr. Tappin continued, "for holding onto, or retaining, the idea of a natural account for the evolution of life from prebiotic beginnings, when the best you seem to be able to give me, at this point, is that the first protocell might have popped into existence despite the calculated odds against such an event happening?

"More specifically, in your scenario about the shifting tides of thermodynamic and kinetic fortune, Dr. Yardley, which, supposedly, have been selecting out, on a consistent basis, certain prebiotic reactions in preference to other possible reactions, you are, in effect, assuming your conclusions. You have assumed that, once upon a time, there was a sequence of thermodynamic and kinetic conditions that had precisely the properties that were necessary to generate and select an extremely long but finite series of reactions that culminated in the life forms we see before us today and that did so through entirely natural means.

"Yet, whenever one begins to examine some of these alleged thermodynamic and kinetic conditions of natural selection with which evolutionary theory is littered, they fall apart before one's eyes. They don't stand up to any kind of careful, reflective consideration.

"The concrete examples that are being offered to the general public by your theory or model, Professor, and some of which we have been exploring during this cross-examination, are intended to serve as a sampling of the kind of thermodynamic and kinetic processes that form the underpinning of evolution's alleged reality and truth. Yet, if these concrete samples don't stand up to examination, then, why should we extend a line of free intellectual credit to evolutionary biologists that permits them to take advantage of the trust that has been invested, at an ever-accelerating rate, in the evolutionary project for the last 140- plus years?"

"As a great evolutionary scientist once said," replied the professor, " 'nothing in biology makes sense except in the light of evolutionary theory'. The theory brings together an incredible wealth of data that cuts across many disciplines such as cosmology, meteorology, geology, hydrology, paleontology, molecular biology, organic chemistry, biochemistry, microbiology, thermodynamics, population genetics, ecology, anthropology, sociobiology and so on.

"Evolutionary theory has great beauty in its dimensions of simplicity, heuristic value, symmetry, integrity, and organizing power. It renders meaningful what would otherwise be inexplicable and disparate pieces of data.

"There is no other scientific account concerning the origins and development of life forms that can compete with, or is a competitor of, modern evolutionary thought. The consensus of the best minds of our time is that irrespective of whatever relatively minor squabbles separate one theoretician from another, or one researcher from another, in broad outline and in its general principles, evolutionary theory has been established beyond all reasonable doubt.

"Focusing on what lends itself to disputation, rather than concentrating on strengths, and engaging in endless rounds of philosophical nitpicking, rather than getting busy with filling in the blanks, are easy, Mr. Tappin. The reality of the matter is, however, that if one rejects evolutionary theory, what is the alternative?"

"The alternative, Professor, is honesty. When you don't know something, admit it, instead of trying to cover up ignorance with a theory that seems to make a lot of sense when viewed from afar, but, when examined from a closer vantage point, one becomes aware of the fact that much of the beauty of this theory is only skin deep.

"I have no doubt there are many, many truths to be found in evolutionary theory, but the problem is, evolution just doesn't seem to be one of them. I see no reason why evolutionary theory should be granted a license to get away with sloppy thinking and presumption when science has never been willing to extend the same latitude to philosophy, religion or mythology.

"Dressing something up in technical language and surrounding it with the pomp of a false rigor, cannot conceal the naked truth. On all too many occasions the evolutionary emperor has little more to wear than the rather threadbare, and all too revealing, cloth of mental presumptions.

"Furthermore, demanding that a critic provide an alternative to evolutionary theory is a little like a prosecutor expecting a defense lawyer in a murder trial to come up with the killer's identity in addition to proving one's client to be innocent. The reality of the matter is, coming up with an alternative to evolutionary theory is not my responsibility since I did not profess to have a solution in the first place."

"Your Honor," intervened Mr. Mayfield, "is there a question in all of this? My esteemed colleague is badgering the witness."

"Yes, Mr. Tappin," noted the judge, "I think this has gone on long enough."

"Very well, Your Honor," the defense counsel acknowledged.

"Before concluding the cross-examination, Dr. Yardley, there are a few more questions that I would like to ask. Please be patient with me for a little longer.

"Let us suppose Professor that the RNA-world hypothesis is true in the sense that whatever ribozymes were necessary to underwrite a fully functioning and self-replicating organism or protocell had, somehow, come into being. How do we explain the transition to a DNA - world in which amino acids are being encoded for rather than nucleotide sequences?

"In other words, although we might assume the kinds of enzymatic functions that ribozymes and proteins perform are similar in character, in a RNA-world the nucleotides in the genome stand for themselves, they don't stand as a code for something beyond themselves as is the case in the DNA-world. In effect, this means a whole new set of nucleotide sequences must be generated when we change over from the RNA-world to the DNA-world, because, in the forms of life with which we are familiar, DNA codes for amino acid sequences not ribozymal nucleotide sequences, and the nucleotide sequences that confer catalytic activity on ribozymes will not necessarily confer catalytic activity on amino acid sequences.

"For instance, if we take the example of the previously discussed IVS RNA ribozymal molecule, then, its 395 nucleotides would have to be divided by 3, in accordance with the requirements of amino acid, formation rules in the genetic code, and this would give a sequence of about 121 amino acids. Not only is this sequence of 121 amino acids unlikely to have the same enzymatic properties as the 395 ribozymic sequence of nucleotides, but there is no guarantee that the 121-amino acid sequence being coded for by the original ribozymal 395 nucleotides, subsequent to the transition to a DNA-world, would have any enzymatic or structural function whatsoever.

"Furthermore, and in an attempt to ensure that what I'm getting at is, hopefully, entirely clear, we will assume there is no trouble in the first stage of transition from the RNA-world to the DNA-world. We are assuming that everything that previously had been stored in RNA, is now being stored by

DNA, so that, during this first stage of transition, DNA can now bring about all of the synthesis of ribozymes that had been handled by RNA-dominated activity in the RNA-world.

"As I see it, the problems that the transition from: an RNA-world, to: a DNA-world tend to pose for evolutionary theory would begin to arise during subsequent stages of the transition process. Even if one assumes continued full ribozymal activity after the first stage of this transition has been completed, how does DNA come to begin coding for amino acid sequences rather than nucleotide sequences, and how does the organism continue to function when the DNA sequences that previously had been coding for ribozymes during the first stage of transition, no longer are doing this?

"Are we to assume Dr. Yardley that yet another incredibly serendipitous event in evolutionary history occurs just in the nick of time? Are we to assume, in other words, that just as each ribozymic nucleotide sequence is lost from the DNA's genetic repository, then, simultaneously, and, yet, quite independently, an encoded nucleotide sequence for a protein comes into being, with precisely the same kind of enzymatic function as the ribozymatic nucleotide that is being lost?"

"Actually, you seem to be assuming," the professor replied, "that the RNA-world hypothesis is the only theoretical game in town. This issue of transition to which you are referring would be a problem only if a DNA-world did in fact arise out of a RNA-world.

"People need to understand that researchers often adopt a given hypothesis on a trial basis and proceed to give it a work out in order to see how it responds under various theoretical and experimental conditions. During this testing period, one tends to finds things about the theory that are appealing as well as features that one dislikes.

"Almost any hypothesis involves tradeoffs between advantages and disadvantages. Researchers might retain a hypothesis because the problems it solves are considered to be more crucial than the problems the hypothesis creates.

"A scientist might develop a working relationship with a hypothesis not because the individual believes the hypothesis is, in some ultimate sense, true, but because the ideas contained in the hypothesis have heuristic qualities that help suggest theoretical possibilities and experiments or help organize and direct thinking in some fruitful manner. A researcher might stay with this kind of hypothesis until something more useful or less problematic or more elegant comes along.

"Although the RNA-world hypothesis solves a number of problems if one adopts it, there are a number of problems that it generates as well. The transition issue to which you alluded earlier is just one of these difficulties.

"In effect, the RNA-world hypothesis requires the genetic wheel, so to speak, to be invented twice. On the first time through, ribozyme - 'nucleotides' are switched over to DNA-'ribozymes' or, one might say, 'dibozymes', while during the second revolution, nucleotides must code for amino acids.

"There are a number of evolutionary researchers and theorists who feel the double-invention aspect of this hypothesis lacks elegance and simplicity. Such people believe that whatever problems might surround the issue of DNA synthesis in Archean era times ... which, remember, was one of the considerations that helped launch the RNA-world hypothesis in the first place ... nonetheless, such unanswered questions, ultimately, might prove to be more conducive to resolution than are some of the difficulties with which we are left in the wake of the RNA-world hypothesis.

"A further possibility is that some sort of hybrid system arose, combining certain elements of both RNA-world and DNA-world scenarios. Conceivably, for example, some of the huge quantities of so-called surplus or junk genetic material that have been discovered in a variety of species, including human beings, and that appears to have no specific function, might have served, at one time, as a kind of laboratory in which various coding schemes were experimented with until something that worked arose, and, gradually, this was introduced into the operations of the cell."

"Is there any evidence," asked the lawyer, "which lends support to the idea that this surplus or junk genetic material might have played a role in helping the first protocell come into existence?"

"Not that I'm aware of," the professor stated. "However, the night is young, so to speak, in the world of evolutionary biology.

"Quantum theory and relativistic physics didn't come into the scientific picture until more than 200 years had elapsed since the Newtonian revolution helped set the stage for much of modern science. Given that only 140-plus years have passed since Darwin helped set the stage for modern biology, I believe many of the questions that you are asking, Mr. Tappin, stand a very good chance of being answered during the next 60 years."

"What you say Dr. Yardley might turn out to be the case," remarked the defense counsel. "Yet, the point that needs to be emphasized is that, at the present moment, evolutionary biology does not have answers to some fundamental questions that affect the plausibility of the origin-of-life problem.

"For instance," posited the lawyer, "if one rejects the RNA-world hypothesis and maintains the first protocell was a DNA based organism, then, presumably, one will have to come up with a plausible account of how the DNA coding system arose. Are we to assume, once again, that randomness has worked its magic and, one fine day, everything suddenly fell into place?

"Moreover, even if we were to allow the randomness assumption to stand, what about the problem of having to explain how a protocell was able to survive sufficiently long for all of this to come together? In fact, this leads to a key issue ... which came first, a working protocell or a working set of genetic instructions?

"Seemingly, Dr. Yardley, no matter that way one goes with these questions, evolutionary biology faces major problems. Wouldn't you agree?"

"One possibility," the professor suggested, "that you might be overlooking is that the problem is not an either-or issue. The idea of co-evolution offers a third alternative.

"Perhaps, the first working protocell joined forces with a developing set of nucleotide sequences, whether RNA or DNA or both, and the two assisted one another in various ways. Perhaps, in the beginning, this mutual assistance only might have been in some minimalist fashion, but, over time, this working relationship might have become refined and more complex."

"Let me see if I understand this, Professor Yardley," replied the defense counsel. "Are you suggesting that, first, a minimally working protocell, somehow, arose spontaneously through a self-assembly process, and, quite apart from this, a mass of nucleic acids, with some kind of minimal or primitive genetic abilities, arose in one of the mini - prebiotic laboratories, and, then somehow, the protocell and the primitive genetic system came together to form a system that became integrated over time such that the genetic instructions that arose in the DNA/RNA system reflected all of the characteristics of the original protocell? Is this what you mean by the idea of co-evolution?"

"Well, I believe," the professor stated, "the idea has suffered somewhat in your translation of it. Nevertheless, in very crude general terms, you have managed to capture some of the spirit of the co-evolution hypothesis?"

"Doesn't this," queried Mr. Tappin, "raise a variation on the same kind of problem that confronts the RNA-world hypothesis? Isn't one asking for the wheel of life to be invented twice?

"More specifically, on the one hand, life is said to be arising in relation to the self-organizing protocell that we are assuming is spontaneously gathering together and assembling all the requisite parts of a cell that, supposedly, are being synthesized in the Archean era environment. On the other hand, life also is arising in the form of a set of genetic blueprints that is capable, among other things, of self- replication.

"In addition, apparently, we are being asked to suppose that the protocell and genetic system: meet; fall in love; join forces; and, somehow, gradually work out their differences over the course of their lifetimes, so that, in the end, their beings have become so inextricably intertwined, not only can't we tell where the protocell begins and the genetic system ends, but the genetic blueprint, somehow, has come to be able to carry an image of the structural architecture of the protocell, much like a lover carries a photo of the beloved. Is this about it, Dr. Yardley?"

"The imagery is somewhat overwrought but serviceable, I suppose, in a very broad sense," replied Dr. Yardley. "In fact, something very similar to the foregoing has been proposed in another context within evolutionary theory.

"In trying to account for how eukaryotic life forms arose from prokaryotic organisms, Lynn Margulis developed what has come to be known as the symbiotic theory of evolution. In this theory, a variety of prokaryotic life forms join together in a symbiotic relationship that, eventually, over time, and through a complicated sequence of increasingly integrated steps of co-evolution, became a new life form -- that is, the eukaryote, whose different internal organelles, such as the mitochondria, might be remnants of what remains of a symbiotic evolutionary history.

"In effect, the coming together of protocells and some sort of primitive system of self-replicating nucleic acids might just be an earlier, cruder version of what could have happened later on with symbiotic co - evolution when the next giant step of evolutionary transformation occurred and the jump from prokaryotic to eukaryotic life was accomplished. One often sees this kind of repeated use of a creative evolutionary strategy take place under different circumstances and at different junctures of evolutionary history."

"I hate to be impolitic about this, Professor," apologized Mr. Tappin, "but I suppose no one has managed to come up with a plausible step-by-step account of how all this is supposed to have happened. Are we dealing here, once again, with that elusive, shadowy and mysterious agent of evolutionary transformation, Mr. Lucky?"

Smiling, Dr. Yardley said: "He's really not such a bad fellow when you get to know him. He's full of strange, wonderful and unexpected things, although his quality of unpredictability can be quite frustrating to deal with for those who are impatient and demand closure on issues right away."

"I have one final question to ask," asserted the lawyer. "As I understand things, there are some 100,000 genes that are encoded in human beings. These genes consist of tens, if not hundreds, of millions of nucleotides."

"Let us assume, Dr. Yardley, as evolutionary theory must, that there is an unbroken chain of genetic lineage reaching back to the original protocell with which this all started. Given this, is there any mechanism in evolutionary biology ... other than the idea of chance, random events -- that can explain how these 100,000 genes, all of which code for different kinds of enzymatic and structural proteins, came into existence? This question is especially important in view of the fact that natural selection can only operate after a gene has arisen, and, therefore, cannot be cited as a cause for the origin of such genes, unless one wishes to argue that quite independently of the function that such a completed gene serves, each and every step of molecular change leading to this gene also was specifically selected by the environment for reasons that we currently can't fathom."

"Relatively recently," Dr. Yardley responded, "the idea of jumping genes or transposable genetic elements ... transposons, for short -- have caused quite a lot of stir in some parts of the evolutionary community. There is a growing body of evidence suggesting these transposable genetic elements might not only move around from one chromosome to another within an individual or a species, but transposons might even be capable of jumping from one species to another.

"Transposable genetic elements seem capable not only of altering the way genes are given expression, but they appear to be capable of becoming inserted into, and integrated with, different genetic systems. If this is the case, then, transposons might constitute a significant medium for potential evolutionary change.

"Although there is still considerable discussion concerning the possible origins of transposons, one hypothesis suggests these transposable genetic elements might be the remnants of viruses that, at one time or another, had integrated some, or all, of their genes into the genome of their hosts. One reason for supposing the virus-transposon theory of origin might have some merit concerns a commonality that seems to be shared by both some viruses and some transposons.

"There are certain viruses possessing a gene for an enzyme known as reverse transcriptase. Essentially, this enzyme permits such viruses to transcribe RNA into DNA.

"Transposons also appear to employ a similar kind of reverse transcriptase mechanism. Sometimes these transposable genetic elements are capable of generating their own enzymes of this sort, and sometimes these transposons will borrow such enzymes from elsewhere.

"There is some evidence indicating transposons often seem to bring about macro mutations. By this, I mean that when jumping genes become inserted into, and integrated with, other genes, then, one tends to observe substantial alterations in the way phenotype, or the total package of physical characteristics of an organism, might manifest itself.

"There even is some evidence being hotly debated that raises the possibility that, under some conditions of environmental stress, certain species of bacteria might enter into a sort of hyper mutable state. In this state, the claim is being made that a variety of mutated offspring are generated in the apparent attempt to overcome, for instance, the species' inability to digest the only available food source in a given environment.

"According to certain researchers, if one, or more, of the mutated offspring happens to come up with the right solution, the colony survives. On the other hand, if there is no such solution forthcoming, then, assuming that the environmental circumstances do not change, the colony dies.

"I don't know, Mr. Tappin, if you would consider this notion of transposable genetic elements to be a non-random element. Nevertheless, there are some intriguing possibilities that arise from this in the context of evolutionary biology."

"Do these transposons," inquired the lawyer, "merely affect how, or if, certain existing genes are given expression, or do the transposons generate new genes in the sense of introducing a totally new enzyme or protein into the phenotypic milieu?"

"The jury is still out on that one," answered the professor. "Even in those bacteria studies that suggest that a new capacity to digest an, heretofore, indigestible nutrient has arisen, no one is entirely sure about what is going on, and there are quite a few scientists who have criticized such studies for insufficient controls as well as for faulty statistical methodology.

"Because we don't know what, if anything, is taking place, we cannot develop any theory about what the possible parameters are that might govern, or regulate, or limit the kinds of evolutionary changes that might be capable of taking place. We might be dealing with a significant force, or a very minor force, for evolutionary change. We just don't know enough at this point."

"Let us suppose" postulated Mr. Tappin "for the sake of argument, that transposable genetic elements have the capacity, on occasion, to code for the introduction of new structural proteins or enzymes. Do genes act in isolation, or do genes tend to work in concert with one another ... not only in the case of gene regulation and expression but also in terms of establishing catabolic and anabolic pathways that involve the action of a number of different enzymes in order to achieve some biologically useful result?"

"Genes tend to presuppose other genes," the professor responded, "since a single protein or enzyme by itself will have, for the most part, a limited capability to bring about any sort of useful biological result. Viruses are a very good example of this since even though they have a few genes, they do not have enough genes to establish, without the help of a host, a means of replication or reproduction. Among other things, they lack: energy storage as well as charge transfer capabilities; ribosomes; enzymes required for the synthesis of various products fundamental to the maintenance of life, and transfer-RNA."

"If," hypothesized the defense counsel, "the bacteria in the experiments to which you referred were able to enter into a hyper mutable state, in your opinion, Dr. Yardley, do you think a whole bunch of new gene are being generated, or do you feel just certain nucleotide sequences were changed in an existing gene that, in one of the mutants, created a gene capable of coding for an enzyme that fit into an existing metabolic pathway?"

"I think the more likely possibility," the professor stated, "is to suppose there was some kind of mechanism for lifting normal regulatory controls on the replication of a limited number of genes, and, possibly, no more than one particular gene. Changing things holus - bolus would not be a good strategy in an organism that already enjoyed evolutionary success. Furthermore, the more changes that are made, the less likely will these changes be capable of being harmonized or integrated with either one another or with the rest of the existing biological system."

"Therefore, can one assume," inquired the lawyer, "that in those cases where an immediate phenotypic change is brought about by the activity of transposable genetic elements, this is because such elements either are affecting the way an existing gene system is being regulated or given expression or because the jumping gene, if it is a new gene in its own right, codes for a protein that has compatibility with an existing pathway?"

"I would say this is a fairly good assumption," the professor affirmed.

"Would one be unreasonable to assume," Mr. Tappin continued, "that in cases where transposable genetic elements don't affect existing gene regulation or expression and do not fit into any of the existing metabolic pathways, then, there is a good chance the new gene would lack the necessary supporting elements of regulation and expression by other genes to be given phenotypic form or manifestation."

"No, I don't believe this would be an unreasonable assumption," the professor indicated.

"Would you agree," asked the defense counsel, "that in order to become phenotypically manifested, the new gene would have to wait for the necessary gene support to arise through the arrival of, say, other kinds of transposable genetic elements? Moreover, would you agree these new arrivals would have to possess the sorts of nucleotide sequence that could code for the regulated and integrated expression of a series of functionally related genes capable of bringing about some sort of coherent phenotypic expression through protein activity?"

"The answer to both of your questions," said the professor "is yes."

"How many genes," wondered the lawyer, "are necessary for an average, integrated unit to be able to be given phenotypic expression."

"This is very hard to say," offered the professor. "The genetic and phenotypic feedback systems affecting gene regulation and expression can be quite complex."

"However, the basic operon model usually consists of, at least, 4 genes. These are known as: an operator gene; a regulatory or repressor gene; an inducer or promoter gene; and, a structural gene that codes for one or more proteins involved in other kinds of biological functioning such as helping to establish some sort of catabolic or anabolic pathway.

"These genes are involved in a set of feed-back relationships. Under certain conditions, a regulatory or repressor gene gets expressed and prevents the operator gene from setting in motion the steps required for the active expression of the different proteins coded for by the structural gene. Under other conditions, an inducer or promoter gene gets activated and produces a protein that can trigger the operator gene to begin operations. In effect, different genes in the operon get turned on and off at various times depending on circumstances."

"On average, Professor, how many nucleotides would be required to encode the information for these four genes?" the lawyer asked.

"This, again, is difficult to say," the professor replied. "Proteins range in size from relatively small ones like insulin that consist of a couple of chains having 21 and 30 amino acid residues, respectively, to monster proteins consisting of multiple chains of amino acid sequences that run into the hundreds of residues per chain.

"Since each amino acid is encoded by a sequence of three nucleotides, even insulin would consist of, at least, 153 nucleotides. In addition, one has to take into consideration there usually are more nucleotides in a gene than what is required to code for just the amino acid sequence.

"For instance, before a particular, transcribed protein code or message leaves a given nucleotide sequence in the form of messenger - RNA, there often are one or more introns, or sequences of nucleotide bases, which get transcribed but are excised or eliminated before the m-RNA, or messenger-RNA, leaves to be translated into proteins at various ribosomal sites."

"Continuing on with the arbitrary nature of this exercise," stipulated the lawyer, "let's set our hypothetical average for a protein, whether structural or enzymatic, at 50 amino acid residues. Would you agree this is likely to be on the low side of the reality of things?"

"Yes," the professor confirmed.

"Let's, also arbitrarily, set the entire number of nucleotide sequences for the four-gene unit at 1000," Mr. Tappin stated. "This figure will take into consideration an assumption that allows for an operator gene, a regulatory/repressor gene, an inducer gene, as well as three proteins within the structural gene component of our hypothetical operon. This figure will permit us to throw in some extra nucleotides -- say, around 100 ... to take care of minor administrative and organizational duties that probably are part of the operon's effective functioning, somewhat like the excised introns to which you referred earlier.

"Now, Dr. Yardley, if we hook up these, admittedly, arbitrary figures and apply them to the 20-25,000 genes that make up the human genome, and that do not take into account the additional 97 percent of surplus/junk DNA of unknown function, then, we are confronted by the following. From the time the last common ancestor arose on: 20 to 25,000 different occasions, an operon of some 1000 nucleotides arose, in order to culminate in a human being.

"These operons would have been involved in establishing a wide variety of new, not previously evolved functions. Thus, as one goes from, say, the time of no immune system to an increasingly complex immune system, a number of new operons would have had to be generated in order to look after the catabolic and anabolic cellular activities that would have to underwrite these biological capabilities.

"Similarly, the emergence of such things as different kinds of hormonal functioning, or embryological activity or enhanced neurophysiological capabilities, would all require the evolutionary biologist to be able to account for the appearance of the new sets of operons that would help manage the regulation and expression of these systems. Indeed, every new organ, organelle or metabolic system presupposes that at some point in evolutionary history, one or more operons somehow came into existence in order to underwrite processes that had not been in existence heretofore.

"Notwithstanding my arbitrary way of alluding to all of this, in order to account for the complexity of a human being, then, on 25,000 different occasions, over a period of some 3.5 to 3.85 billion years, a specific sequence of 1000-nucleotides was, somehow, selected from a total of 41000 possible nucleotide combinations. On average, I believe this means that once every, very roughly, 100,000 years, this search for a functional sequence of 1000-nucleotides with phenotypic survival or selective value must be solved in the midst of 41000 possibilities.

"One could work out, I'm sure, how many mutational experiments would be necessary, on a moment to moment basis, over the course of 100,000 years to explore even a small fraction of the total combinatory possibilities that are available with respect to one operon, consisting of 4 genes and 1000 nucleotides. I will, as they say, leave it as a homework exercise and, I believe anyone who cares to perform the calculations will conclude this seems to be carrying the idea of hyper mutability beyond the realms of believability.

"My question, Dr. Yardley, is this. How exactly does the idea of transposable genetic elements reduce the element of randomness that seems to saturate the foregoing figures? That is, how does the notion of the transposon, as an alleged agent of evolutionary change, but whose origins are lost in the swirling mists of chance events, permit one and all to see through the mysterious shadows cast by the impenetrable nature of randomness and understand, as you previously suggested is the case, that beyond any reasonable doubt, evolutionary theory is, indeed, true?"

While the professor was reflecting on the question, Mr. Tappin held up his hand in a sort of halting motion. The lawyer said: "To show you what a sporting fellow I am, Dr. Yardley, and to indicate what I think about the idea of chance, I'm going to give you a chance and withdraw the question.

"Your Honor, my cross-examination of this witness has concluded.

"Mr. Mayfield," indicated Judge Arnsberger, "you may call your next witness.

"Your Honor," replied the prosecuting attorney, "we have no further witnesses scheduled to appear. In the matter of the People versus Wayne Robert Corrigan, the prosecution rests."

"Are you prepared to proceed at this time with your first witness for the defense, Mr. Tappin?" inquired the judge.

"Your Honor," he responded, "the defense is prepared to rest its case. Furthermore, if it pleases the court, we would like to move that the charges against our client, Mr. Corrigan, be dropped for lack of sufficient evidence."

"The motion doesn't please the court," Judge Arnsberger asserted. "Whenever possible, Mr. Tappin, I prefer to leave judgments concerning matters before the court in the hands of the people. I believe this is why we have a jury system. They are the fact finders and ones who weigh the evidence, not the judge. Motion denied."

"Are counsels for the prosecution and defense ready for summation?" the judge asked.

"The people are ready, Your Honor," Mr. Mayfield stated.

"The defense, also, is prepared, at this time, to proceed with closing remarks, Your Honor," said Mr. Tappin.

"Mr. Mayfield," Judge Arnsberger announced, "the floor is yours."

Closing Arguments

The prosecutor rose from behind his table and approached the jurors with a smile on his face. When he was a few feet away from the end of the jury area nearest his table, he stopped.

He surveyed the jurors for a few seconds, and, then, he walked to a mid-point several feet removed from the jurors. Slowly at first, but quickly picking up a little speed, his words began to flow.

"Ladies and gentlemen of the jury, you have been very patient and attentive during the last several days of testimony and cross - examination. However, in many ways, your most important task lies ahead of you.

"Now, you not only have the responsibility of making sense of a great deal of information and technical argument, but you also have a duty to come to a judgment about the guilt or innocence of a fellow human being. Such activities can neither be taken lightly by, nor can they rest lightly with, any of you.

"I, as the prosecuting attorney, also have duties and responsibilities. Both during the trial, as well as currently, during these closing remarks, I have had the job of putting forth a case, within the limits imposed upon all participants at the outset of this trial, that would provide you, the members of the jury, with sufficient reason to come to the only conclusion that I believe reflects the totality of evidence ... namely, that Wayne Robert Corrigan is guilty of teaching material that contravenes well-established principles of evolutionary theory, as well as, of scientific methodology.

"During the evidential portions of the trial, I have attempted to fulfill my task in several ways. First, you have been supplied with materials that constitute the written part of Mr. Corrigan's curriculum, and I believe these materials speak for themselves.

"Secondly, I sought, and secured, the co-operation of one of the world's leading evolutionary biologists, Dr. Alan Ross Yardley. For several days, we all have been enjoying the benefits of being able to listen to an eminently qualified expert talk eloquently, precisely and movingly about his discipline.

"Nonetheless, without, in any way, wishing to diminish the quality or value of Dr. Yardley's participation in these proceedings, there are two points, ladies and gentlemen of the jury, that I would like to bring to your attention. Moreover, these are points with which, I am completely certain, Dr. Yardley would concur were he to be asked his opinion on the matter.

"First, Dr. Yardley is but one individual, among thousands of very qualified and gifted professional scientists, who has been called to give testimony in this case. With all due respect to Dr. Yardley's eminence as a scholar, many, many people could have been asked to give testimony, and each of them would have been able to provide the same kind of standard of excellence and expertise as has Dr. Yardley.

"They could do this because they are part of a community of scientists and researchers who have dedicated their lives and talents to the pursuit of what can be known by human beings on the basis of a disciplined, rigorous and methodical application of reason to human experience in the context of a physical and material world. Any of these researchers and scientists could have substituted for Dr. Yardley because they all are contributors to, as well as inheritors of, the treasury of accumulated knowledge and wisdom that has been struggled for, through tireless efforts, in the unchartered and, at times, dangerous territories at the frontiers of human understanding.

"One of the reasons these sorts of struggle can be dangerous is because when knowledge and wisdom come, lives that are ruled by ignorance, superstition, and habit are threatened. Under such circumstances, historically, the tendency of vested interests that feel threatened is to be reactionary and strike out in harmful ways at those who would have the temerity to throw back the curtains of conceptual darkness that are preventing light from coming into the life of the mind.

"We owe people like Dr. Yardley a debt of gratitude for the way they have stood their intellectual and moral ground, for more than one hundred and forty years, against people, like Mr. Corrigan, who seek to hold on to the familiar and convenient at the expense of the truth. Courageous individuals from: Charles Darwin, to: Allan Yardley have risked much in order to help humanity to transcend its tendency to become locked into non-productive patterns of intellectual inertia and lethargy."

Mr. Mayfield stepped back a few paces from the place where he had been standing. He began to walk slowly, back and forth, in front of the jury area, using his arms to help give animated expression and emphasis to his words.

"The other point, ladies and gentlemen of the jury, to which I wish to draw your attention, again with no wish to cast aspersions upon the quality of Dr. Yardley's wonderful testimony about, and defense of, evolutionary theory, concerns the following. I don't believe I can adequately stress the importance of understanding that the evidence which was forthcoming from Dr. Yardley during direct examination and cross-examination is but a tiny subset of the amount of information, knowledge, data, experiments, analysis and reflection that bears upon the issue of evolution.

"When one brings together, in dynamic juxtaposition, firstly, the dedicated expertise of the community of scientists and researchers who were the focus of my first point, as well as, secondly, the wealth of understanding concerning evolution that has matured over the last century and a half, which was the focus of my second point, then, one cannot help being deeply affected by the strength, depth, richness and sophistication of evolutionary thought. Those individuals, who are among the best, the brightest, the most skeptical, the most rigorously analytical and demanding minds in the history of humankind, have established an inter-subjective consensus concerning the truths at the heart of evolutionary theory.

"I suppose one can forgive the fact there are people who, perhaps as a result of an inadequate or poor quality of education, ask the question: if evolutionary theory is so true, why is it still only a theory? Why don't we raise the epistemological status of evolutionary thought?

"Ironically, the reason for retaining the moniker of 'theory' actually has more to do with the integrity of the scientific process than it does with any presumed, tacit admission there is something inherently wrong with this discipline. Although the observational data, facts, experimental results, principles and laws that form the substantive foundations of evolution have been established scientifically and are agreed upon by the community of evolutionary researchers, both past and present, nevertheless, and evolutionary biologists are the first to admit this, there still is much work that needs to be done in order to discover the many things that continue to elude our understanding at the present time.

"None of these unknowns is expected to undermine anything that has been established, and agreed upon, to date. If anything, when these unknowns are discovered and added to the treasure house of our knowledge concerning the process of evolution, they merely will deepen our appreciation and understanding of the complexity and intricacy of nature as it manifests itself through, among other things, evolutionary phenomena.

"Scientists know that as our understanding changes, modifications have to be made in the conceptual framework, model or theory that is being held up as a mirror, of sorts, to natural events. Among scientists there is an awareness of the difference between our understanding of something and the reality, whatever this term might ultimately mean, of that to which our theories are attempting to make identifying reference through descriptions, explanations and so on.

"As Dr. Yardley indicated at one point during cross-examination, science is a work in progress. One can acknowledge this unavoidable truth and, nonetheless, maintain that the conceptual changes that are inevitable are played against a background of fundamental truths whose essence does not change even if the vocabulary through which they are given expression might change with time.

"Quantum and relativistic theory did not alter the truths that had been established previously by science. Rather, these revolutions changed the way we understood, and made use of, what already had been established and known, and, in addition, these ways of thinking helped bring about tremendous contributions to, and the growth of, the repository of human knowledge.

"What disappeared from the intellectual scene in the wake of these revolutions as they passed through the physics and scientific communities, were the ideas, hypotheses, conjectures, theories and models that were rooted in the untenable ways of organizing and interpreting the knowns of science. These revolutions showed more viable, more heuristically valuable, more elegantly fundamental, and more beautiful ways of organizing the knowns of science.

"Evolutionary thought is in the process of effecting the same kind of changes in a variety of biological and associated disciplines. Yet, there are many social, religious, political and philosophical forces that are attempting to resist, and interfere with, efforts to proceed with the exploration, and expanding, of the horizons of human understanding.

"We need to be very clear in our focus on these matters. We need to understand that if people, like the defendant, are permitted to teach anything they like ... no matter how much it might fly in the face of well-established scientific facts, principles and knowledge ... then, we are doing a great disservice to our children and future generations of children.

"To permit the Wrong-way-Corrigans of the world to ply their trade in our classrooms will lead to the development of only confusion, ignorance, and scientific illiteracy among students. To allow individuals such as Mr. Corrigan to indoctrinate children with a dogmatism that can only corrupt and diminish human potential, is to abandon the fiduciary responsibility to humanity that each and every one of us has by virtue of coming into this world as human beings.

"People such as Wayne Corrigan wish to interpose themselves between their students and the community of scientists and say: I know better than these experts and professionals who have dedicated their lives to mastering their disciplines. People like Wayne Corrigan have dropped the gauntlet before society and belligerently pronounced: I refuse to pass on the legacy of understanding and knowledge that has been bequeathed to students by the researchers of the scientific community.

"The Wayne Corrigans who live among us have a tendency to envision themselves as courageous individuals who are fighting the lonely battle against the forces of repression that, in this case, are allegedly being perpetrated by science, in general, and evolutionary theory, in particular. In reality, all too frequently, these individuals are merely caught up in their own megalomania and wish to entangle everyone else, and especially vulnerable students, in their delusions as well.

"Individuals like Wayne Corrigan have no viable alternatives to offer to evolutionary theory. Instead, they prefer for all of humanity to sit idly about, twiddling its collective thumbs, and saying: but you evolutionary scientists haven't proved this relatively minor point or you haven't demonstrated that minor point.

"They ignore the scope, power, value, beauty, elegance, richness, and productive capacity of evolutionary thought, and, consequently, these individuals wish to jettison these important aspects of our cultural manifest, as so much jetsam, in order to save the conceptual ship that they believe is in imminent danger of floundering amidst the rocks of moral turpitude that they associate with scientific activity.

"They are like Don Quixote's evil twins who are flailing away at imaginary windmills but who do so for something other than noble ... albeit, rather excessively and romantically misguided ... idealistic purposes. Instead, they won't be happy until everyone thinks in the same profoundly limited and superficial fashion as they do.

"Ladies and gentlemen of the jury, the choices before us are fundamental in character. We can proceed into the unknown with people of rigorous and methodical dedication like Dr. Yardley who might not have all the answers but who are committed to finding them, or we can proceed into the future by returning to a regressive and dogmatic past like people such as Mr. Corrigan who believe they have all the answers, and, therefore, there is nothing left to discover.

"I have confidence in your ability to make the correct and courageous choice and, consequently, I believe you will endorse the People's belief that Mr. Corrigan is, indeed, guilty of teaching material that conflicts with established principles of both science and evolutionary theory. I beseech you to find Mr. Corrigan guilty and establish a precedent for which history and our children will be eternally in your debt.

"Ladies and gentlemen of the jury, I wish to thank you, again, for your time and consideration. I know you will faithfully fulfill your duties and responsibilities with respect to the matter that is before this court."

Mr. Mayfield nodded his acknowledgement of thanks and returned to his seat. As he sat down, he poured himself a glass of water and began to drink from it.

"All right, Mr. Tappin," Judge Arnsberger said, "you may offer your summation."

The defense counsel rose from his chair and began speaking almost as soon as he was standing. He continued to speak as he gradually made his way to the general area of the jury.

"Mr. Mayfield would have us all believe the issue that is to be decided by you, the members of the jury, is whether or not Mr. Corrigan has taught students in a manner that is in conflict with the principles of evolutionary theory and scientific methodology. One of the problems with this perspective of the prosecuting attorney is that no one, least of all him, has been able to demonstrate just which specific principles of either evolutionary theory or scientific methodology are, allegedly, being contravened by Mr. Corrigan.

| Origin of Life |

"When you look through the curriculum materials that have been introduced as People's Exhibit 'A', you will find that Mr. Corrigan is advocating nothing except the following. An individual should not accept conclusions, scientific or otherwise, until there is a demonstrable chain of evidence that is capable of lending plausible support to the claimed link between the premises of an argument and the conclusions that are said to follow from those premises.

"In addition, you will find in those curriculum materials that a wide number of methods have been developed and elaborated that are designed to help students engage evidential claims from a variety of analytical, reflective, contemplative, experiential and interpretive vantage points. Those curriculum materials, in fact, constitute, and I'm sure you will agree, once you have had an opportunity to examine those materials, a rather intense investigation into the varieties, possibilities, and problems of methodology.

"Difficulties arise, however, at least as far as Mr. Mayfield is concerned, because Mr. Corrigan has the audacity to suggest scientific methodology is not the be all, and end all, of epistemology. Mr. Corrigan, in other words, is questioning the legitimacy of the tendency of many scientists to arrogate to themselves the role of being final arbiters in all matters involving analysis of, critical reflections on, and interpretations about the meaning, value, significance, tenability, truth and rigor of scientific statements.

"For someone to say that such-and-such is what scientists do or that so-and-so is what scientists have agreed upon, is one thing. To make the claim that because this is what scientists do and this is what scientists agree upon, then, therefore, ... especially someone who has not gone through the validation and accreditation process of professional science ... who is critical of what scientists do or say must be dismissed as a fanatic, is an entirely different matter.

"Science is but one approach to dealing with, and understanding, various facets of the phenomenon of lived experience, and, quite frankly, it is an extremely limited way of trying to understand the breadth and depth of what is entailed by being human. Science is but one kind of activity among many possibilities such as law, art, music, literature, philosophy, religion and mysticism that are all capable of deepening human awareness of the many, many factors that can affect how we perceive, interpret, value and act upon experience, including scientific experience.

"Science, in fact, knows little or nothing about a variety of tools that it presupposes in all of its endeavors. More specifically, science knows virtually nothing at all about the processes of consciousness, creativity, thought, insight, interpretation, or understanding that frame, color, orient and shape every cubic nanometer of scientific activity.

"Moreover, individual scientists are as vulnerable to bias, prejudice, error, distortion, and dogmatism as any other group of people. In addition, collectively, scientists have demonstrated throughout their illustrious history that just because the generality of scientists agree upon something is no guarantee of the truth of whatever it might be on which agreement has been reached.

"Many of the most vociferous opponents of Galileo, Copernicus, Kepler, Newton, Darwin, Planck, Einstein and so on, were eminent and respected scientists of their days. Scientific revolutions are called this because of the vast upheaval that they introduced into the thinking, methods, ideas, practices and understanding of, among others, scientists ... many of whom were extremely resistant to what was being proposed by a given revolution in science.

"The activities of scientists that help shape the nature of science have their share of politics, pettiness, lack of vision, inertia and blindness. Furthermore, scientists and science are not autonomous entities that are independent of the cultural, social, political, economic and religious milieu in which they operate.

"Scientists, each according to her or his ability, might be dedicated to truth. Yet, many of them also tend to become entangled in a variety of: associations, networks, vested interests, processes of marginalization, and value judgment s that frequently have fundamental effects on who and what gets funded, published, hired, and taught.

"Mr. Corrigan believes in teaching his students to be skeptical of, but open to, a variety of possibilities. He encourages his students to be: analytical, reflective, contemplative, critical, fair, honest, creative, eclectic, practical, idealistic, thorough, experimental, as well as dispassionate but committed.

"He wants his students to become aware of their own assumptions, prejudices, and biases. He tries to help his students come to a fundamental realization that the dynamics of perception and interpretation are shaped and colored by a lot of individual, professional, cultural, historical and philosophical factors.

"Mr. Corrigan is interested in trying to instill in his students a deep awe, respect and love for the pursuit of truth and understanding. He does whatever he can to inspire his students to work toward acquiring a sense of joy and excitement concerning the exploration of human existence.

"He teaches his students to take the issues of methodology seriously and not to leave the subject matter in the classroom. He wants his students to understand that a judicious methodology has implications for self, life, meaning, values, and community.

"If any of this is in conflict with the principles of science, then, perhaps, the time has come to get rid of those aspects of science that are in conflict with the kinds of thing that Mr. Corrigan is attempting to teach his students. If anything, he has run into difficulties because he has held up a mirror to the way evolutionary scientists go about plying their trade and questioned whether such practices constitute satisfactory epistemology, let alone sound science.

"What does it mean to say a given chain of evidence is a plausible one? The curriculum materials, that constitute People's Exhibit 'A', attempt to explain why Mr. Corrigan believes there are major, not minor, problems with the chain of so-called evidence that is cited by many scientists and biologists as justification for the conclusion that natural evolutionary processes adequately account for, among other things, the origin-of-life.

"During cross-examination a very extensive sampling of evolutionary thought has been investigated in some detail. We have taken a look at: cosmological theories concerning the origin of Earth; asteroid bombardments; interstellar dust clouds; interplanetary dust particles; carbonaceous chondrites; differentiation of the Earth's magnetic core; ocean formation; atmospheres of various kinds of reducing and non-reducing composition; ocean-vaporizing impacts; photic-zone vaporizations; interpretation of Carbon [12] and [13] isotopes in the Isua rock formation; the faint early sun paradox; run-away greenhouse effects; ocean pH values; ultraviolet radiation; shock-wave synthesis; processes of photolysis, hydrolysis, and pyrolysis; possible synthetic pathways for hydrogen cyanide, formaldehyde, amino acids, ribose sugars, nucleic bases, phosphates, fatty acids, and phospholipids; the nature of membrane functioning; porphyrin pigments; issues of chirality or handedness; cross-bonding potential in prebiotic condensation reactions; Strecker synthesis in the Archean era ocean; Fischer-Tropsch mechanisms; formose reactions; alleged simulation experiments; problems of polymerization involving proteins, DNA, and RNA; issues of replication; the RNA-world hypothesis; ribozymes; natural selection; evolutionary pressure; transposable genetic elements; the possible role of random events; origins of the genetic code; protocell formation; co-evolution, and the operon model.

"At each and every stage of our investigation there were major, unresolved questions concerning the tenability and plausibility of the evolutionary model. There is no consistent, rigorous chain of evidence that starts from first principles concerning known facts about the natural processes of cosmology, geology, hydrology, meteorology, thermodynamics, inorganic chemistry, or organic chemistry, and which permits one to see that, in principle, if not in broad detail, there is a plausible path that is capable of leading any reasonable individual to understand how life originated through purely natural processes under what we believe to have been Archean era conditions.

"This is not a matter of two people looking at the same glass of water, and one person seeing it as being half empty, while the other individual perceives it to be half full. This is a matter of too many assumptions, problems, questions, ambiguities, uncertainties, unresolved dilemmas, and unbridgeable, at least at this time, conceptual chasms.

"Now, an evolutionary researcher might look at the mounds of data, experimental results, technical models, or mathematical formulae and believe this is all great science. In reality, however, this kind of science not only fails to demonstrate the validity, or even tenability, of a plausible evolutionary account concerning the origin-of-life, but anyone, given what is currently known and understood, who should try to claim that the available evidence supports, beyond a reasonable doubt, a natural account of the origin-of-life, is engaging in both bad science as well as terrible epistemology.

"There have been a great many books, articles and so on that have been written by scientists and others who have been severely critical, and rightly so in my opinion, of the attempt by creationists to try to pass creationism off as a science. People degrade the magnificence of creation by reducing it to the very limited, narrow preoccupations of the world of physical, material science.

"On the other hand, almost no criticism has been directed toward scientists for attempting to pass off evolutionary theory as a disguised form of faith. Ultimately, however, evolutionary accounts of the origin-of-life require one to have faith in the great deities of chance and assumption.

"The deities of chance and assumption render all things possible. Whatever your theoretical problems might be, these deities can resolve them.

"There is no process, reaction, event, or possibility for which provisions cannot be forthcoming from the infinite powers of chance and assumption. Whatever theoretical rivers need to be forded, or whatever conceptual mountains must be scaled, or whatever evidential chasms need to be bridged, the deities of assumption of chance are present and waiting for the faithful to call out in supplication.

"The miraculous, the inexplicable, the amazing, and the incredible become the commonplace by the grace of the holy writ of the law of large numbers and the givens of assumption. Seek, and you shall find; ask, and it shall be given to you; knock, and all doors will become open to you.

"Little is required of you to adopt this faith. All you need to do is take advantage of the opportunities that chance provides and assume everything turns out okay.

"The faith is simplicity itself. For those who are prepared to submit, beauty and meaning shall flow into their lives like manna from heaven.

"The litanies are easy to learn. Just say: 'if', 'given', 'possibly', 'conceivably', 'assuming', 'probably', 'theoretically', and 'plausibly', and all manner of things will be added unto you.

"I would love to play golf with evolutionists. I can see myself standing at the 18th hole at Augusta and saying: 'If we assume that I hit this just right and, then, get a few, chance, lucky bounces in my favor, I believe I could hole out ... what do you think?' I'm sure they would treat it as a 'gimme'.

"We could play a round of golf and never leave the clubhouse. All we would have to do is assume our way through eighteen holes.

"We could shoot 18, or less, every time out. Who could argue with us since we would have consensual validation on our side?

"Of course, people might begin to suspect something was amiss with our consistently, incredibly low scores. They even might want to make a federal case out of it.

"If this were to happen, however, neither I nor my evolutionary golfing buddies would have anything to fear. We would just get Dr. Yardley to testify on our behalf as to how plausible our account was despite the numerous improbabilities, problems and questions surrounding our theory of what was happening out on the golf course.

"Mr. Corrigan has made the mistake of stepping on the toes of those who are deeply committed to their faith. These zealots take umbrage with anyone who would ridicule their faith as being merely a myth told to impressionable children in order to help the youngsters make sense of, and feel at home in, a bewildering, mysterious, and, sometimes, frightening universe.

"Their deities are jealous gods who do not tolerate worship at any other alter. The guardians of the faith and the keepers of the ark of the chance covenant with assumption are quite certain that all those who do not bow down to the idol of evolution will surely be condemned to an eternal doom in the outer darkness where there will be much wailing and gnashing of teeth.

"Ladies and gentlemen of the jury, you have an opportunity to stop this nonsense. Someday, somebody might come along and be able to demonstrate, in a plausible fashion, and beyond any reasonable doubt, that there is an unbroken chain of evidence that links first principles of science with a purely natural account of the origin-of-life."

"Today is not such a day, and no one, so far, has provided anything remotely approaching such a plausible chain of evidence. Consequently, we need to demand that the classrooms of our nation be made safe for the teaching of science uncontaminated by matters of faith.

"The teaching of biology is a wonderful thing. However, including evolutionary theory as part of the biology curriculum is a violation of the Constitution's separation clause between the state and matters of faith.

"I ask you to find Mr. Corrigan innocent of all charges. I ask you to make the sort of judgment on this issue before the court that you know in your heart is the right thing to do. Thank you!"

www.ingramcontent.com/pod-product-compliance
Lightning Source LLC
Chambersburg PA
CBHW020638220526
45464CB00001B/206